W0115365

POSTMOBILIZATION TRAINING RESOURCE REQUIREMENTS

Army National Guard

Heavy Enhanced Brigades

THOMAS F. LIPPIATT
JAMES C. CROWLEY
PATRICIA K. DEY
JERRY M. SOLLINGER

Prepared for the
United States Army

Arroyo Center

RAND

Approved for public release; distribution unlimited

This project analyzes the resources required to conduct the postmobilization training of enhanced heavy brigades of the Army National Guard. The project was sponsored by United States Army Forces Command, and it was carried out in the Arroyo Center's Manpower and Training Program. The Arroyo Center is a federally funded research and development center sponsored by the United States Army. The report should interest those concerned with the reserve components, their training, or the mobilization and deployment requirements of U.S. forces.

CONTENTS

FIGURES

BACKGROUND AND PURPOSE OF STUDY

The Army National Guard (ARNG) combat brigades that were called up for Operation Desert Storm took longer to get ready than many people had anticipated.[1] As a result, some steps were taken to improve the peacetime training of enhanced National Guard brigades, with an eye to improving their ability to respond to short-notice contingencies. However, postmobilization training remains an issue, particularly in light of the substantial force reductions made since Desert Storm. Active-duty combat units played a large role in helping the reserve component brigades prepare for deployment to the Persian Gulf, and it is likely that few if any will be available to provide such support in the future.

The Arroyo Center was asked to determine whether there are adequate resources—sites, training and support personnel, and opposing forces—to prepare the seven enhanced heavy National Guard brigades for deployment.[2] To make that determination, we developed a detailed training model that allows us to quantify the resources required to train a single heavy brigade and its support elements. We also examined different alternatives for implementing the model, that is, different numbers of sites.

THE TRAINING MODEL

The training model contains the events necessary for an enhanced ARNG heavy brigade to carry out the most difficult mission: engage in combat shortly after deployment. To prepare the unit for this mission, we patterned the training model on the policies and practices of the active component. That is, we applied to our model the policies and practices active component table of organi-

---

[1] General Accounting Office (1991), Department of the Army Inspector General (DAIG) (1991).

[2] There are 15 enhanced brigades. Besides the 7 heavy brigades there are 7 light brigades and an Armored Cavalry Regiment (ACR). We examined only the heavy brigades.

zation and equipment (TOE) units use in preparing themselves for combat. The requirements for battalion force-on-force maneuver and brigade-level operations have the greatest effect on resources, creating the need for an opposing force to maneuver against the brigade and the attendant maneuver areas. Our goal of carrying out the training in the shortest possible time also affected resources, because we employed parallel training, that is, all units are training all of the time. This technique requires more trainers, training support, and training facilities than a sequential approach. Reducing the resources would either lengthen the time or lower the quality of the training.

The model requires approximately 100 days to execute.[3] This period includes a fairly intense training schedule of about 80 days, plus time for preparation, movement to the training site, and maintenance. More capable units could compress this schedule somewhat; less capable units will require longer. Furthermore, the situation in theater at the time of mobilization will affect the training time.

We do not imply that the model we have developed is the recommended solution for postmobilization training of National Guard brigades. We developed our model as a way of quantifying resources. Obviously, there are numerous alternative approaches. However, the application of the active-duty paradigm seemed reasonable for units being prepared to enter combat. If the mission were different—say, replacing units in Europe that had deployed to the combat theater—then a very different model would apply, and a different set of resources would be required.

ASSUMPTIONS

The training model rests on several assumptions, some of which are optimistic. These assumptions must be met for the model to produce a brigade trained to enter combat in the time allotted. Some key assumptions are that the brigade will be C-1 in equipment and personnel 18 days after it is mobilized (M+18), that the brigade will be at the training levels reached by the better round-out brigades during 1992 and 1993, that trainers and an opposing force (OPFOR) will be organized and ready to begin training by M+18, and that the postmobilization program can be logistically supported.

---

[3]This figure is RAND's estimate, not official DoD policy. DoD statements have expressed the goal of having the brigades ready to deploy within 90 days after call-up (Aspin, 1993, p. 94; Lee, 1995). Our training model requires slightly more time, based on detailed analysis of the steps needed to prepare brigades for combat missions.

RESULTS

Our analysis indicates that sufficient resources exist to run three brigade training sites with three gunnery sites for preliminary training. There are some trainer shortfalls, but they are small enough that the needed individuals could be provided by expedited individual replacement mechanisms.

Although it is possible to operate three brigade-level sites, the resource bill is substantial, and there are risks that the training objectives and timelines will not be met. To conduct training at the six sites mentioned above requires the personnel shown in Table S.1.

In general, the trainers and training management personnel are drawn from active component sources, such as trainers at the National Training Center and active-duty personnel who provide training support to the reserve components during peacetime. Training support, installation augmentation personnel, and the opposing force (OPFOR) are drawn from reserve component units.

Operating three brigade sites with three gunnery training sites will generate three trained brigades in 102 days and six brigades in 156 days. Operating fewer sites generates trained units more slowly but requires less pre- and postmobilization resources and invites less risk. One brigade site with two company training sites produces three brigades in 172 days and six in 262 days, and two brigade sites with two gunnery training sites can deliver three and six brigades in 159 and 226 days respectively. Our analysis indicates that it is not feasible to operate more than three brigade-level sites. The lack of experienced personnel to support training imposes the most restrictive constraint.

We make no recommendation on how many sites to operate. That decision must result from the policymaker's analysis of tradeoffs along three dimensions: risk, resources, and force-generation rates. Risk refers to training quality

Table S.1

Personnel Requirement for Three Brigade and Three Gunnery Sites

Type	Number
Trainers and training management	1,859
Training support	
Lanes and ranges	625
Field support to trainers	198
Installation augmentation	7,809
OPFOR	11,372
Total	21,863

and timelines. As more sites are staffed, it becomes more difficult to meet the training model's assumptions, the available pool of experienced trainers spreads more thinly across sites, and the more likely it is that quality of training will decline. Also, as training expertise is stretched across more sites, the opportunity increases for problems to arise. The training model contains little margin for delay, and any significant problems could interrupt the schedule, resulting in a slower force-generation rate. Obviously, the more sites operated, the more resources are required, but the faster trained units can be produced.

IMPLICATIONS

The training model and the resulting resource requirements have a number of implications. The model produces a brigade trained to enter combat. Other potential missions, such as backfilling an active unit that deploys from Europe, would require considerably less time, although that mission would not decrease the resources substantially. Furthermore, the model assumes the brigades, when mobilized, have achieved a level of readiness that equals that of the better brigades observed in 1992 and 1993. Some brigades currently cannot meet this standard, in large part because of difficulties in sustaining personnel readiness. Thus, some of the units will have difficulty meeting the model's timelines.

Furthermore, the model requires a trained OPFOR, and this requirement has implications for the reserve components. Our analysis calls for the reserve components to meet the bulk of this OPFOR requirement, which means that they must train as an OPFOR in peacetime. A skilled OPFOR is an important part of the model, because it is required to identify the training weaknesses of the unit being trained and not inhibit training. To achieve the necessary level of proficiency as an OPFOR, reserve component units might require training time and other resources (e.g., equipment, increased active-duty support) beyond levels they currently have.

Some of the implications have significance for the entire Army. One of the most significant is that considerable premobilization planning and preparation must take place if any of the alternatives described are to occur smoothly early in the postmobilization period. Little planning and preparation at the level of detail needed for this is now ongoing. Furthermore, these activities would represent an additional and difficult requirement in light of current peacetime resources.

Second, the model implies that the training sites will require substantial logistics support in terms of spare parts and ammunition to sustain the intensive training. Those resources will be required at a time when other, higher-priority units are also preparing for deployment. Should the Army not be able to accommodate the surge in demand, the time required to train the brigades could lengthen.

ACKNOWLEDGMENTS

We are indebted to many Army officers and Department of the Army civilians who provided insights, access to information, and suggestions that greatly assisted this study's efforts. We owe particular thanks to Brigadier General Walter Mather, Colonel John May, Jr., Colonel James Mowery, Colonel Harrison Lobdell, III, Lieutenant Colonel Joseph Hummel, Lieutenant Colonel James Muhl, Mr. Danny Springer, Mr. Fredrick Stritzinger, Mr. Robert Wade, Mr. Herbert Payne, and Mr. Dewitt Houser at Headquarters FORSCOM; Colonel Randy Gordon at the office of the Assistant Secretary of the Army (Manpower and Reserve Affairs); Colonel Frank Cook, Colonel Eric Braman, and Lieutenant Colonel (P) F. Wilson Myers at the Army National Guard Bureau; Colonel Kenneth Smith, the Senior Advisor for the State of Washington and former chief of the 6th Army ORE Team; Colonel Richard Geier, Captain Andy Cox, and Captain Gutierrez at the 2nd Reserve Training Brigade; Brigadier General William Wallace, Brigadier General James P. O'Neal, Colonel Terry Tucker, Colonel Patrick Lamar, Lieutenant Colonel Phil Laser, Colonel John Anderson, Colonel Ronald Thomas, Lieutenant Colonel Harry Leiferman, Mr. Greg Emmerling, and Mr. Rick Travis, all at or formerly at the National Training Center; Colonel Thomas Kelley and Captain John Garrett at Readiness Group Fort Lewis; Lieutenant Colonel Tony Giordino and Mr. Angel Castel at 1st U.S. Army; Lieutenant Colonel DeLong at 6th Army; Mr. Doug Melton at Fort Carson; Mr. George Bankston at Fort Bliss; Lieutenant Colonel Andy Raymond, Mr. Tom Meeks, and Mr. John McGrafton at Fort Lewis; and Mr. Cliff Watson at Fort Stewart.

AC	Active Component
ACR	Armored Cavalry Regiment
ADA	Air Defense Artillery
AGR	Active Guard/Reserve
ARNG	Army National Guard
ARTEP	Army Training and Evaluation Plan
AT	Annual Training
BCBST	Brigade Command Battle Staff Training
BCE	Bradley Crew Examiner
BCPC	Bradley Crew Proficiency Course
BCST	Battle Command Staff Training
BCSTB	Battle Command and Staff Training Brigade
Bde	Brigade
BFV	Bradley Fighting Vehicle
BGST	Bradley Gunnery Skills Test
BSFV	Bradley Stinger Fighting Vehicle
C2	Command and Control
CALFEX	Combined Arms Live-Fire Exercise
CEV	Combat Engineer Vehicle

CFP   Contingency Force Pool

CFX   Command Field Exercise

CMTC   Combat Maneuver Training Center

COFT   Conduct of Fire Trainer

CONUS   Continental United States

CONUSA   Continental United States Army

CPX   Command Post Exercise

CS   Combat Support

CSS   Combat Service Support

CTC   Combat Training Center

CTT   Common Task Training

DA   Department of the Army

EXEVAL   External Evaluation

FDC   Fire Direction Center

FEB   Field Exercise Brigade

FIST   Fire Support Team

FORSCOM   U.S. Army Forces Command

FSB   Forward Support Battalion

FSO   Fire Support Officer

FSS   Fire Support Section

FTG   Field Training Group

FTX   Field Training Exercise

GFRE   Ground Forces Readiness Enhancement

IDT   Inactive Duty Training

IFV   Infantry Fighting Vehicle

IRR   Individual Ready Reserve

JRTC	Joint Readiness Training Center
MAPEX	Map Exercise
MAT	Mobilization Assistance Team
MC	Mission Capable
METL	Mission Essential Task List
METT-T	Mission, Enemy, Terrain, Troops, Time Available
MILES	Multiple Integrated Laser Engagement System
MOS	Military Occupational Specialty
MPRC	Multipurpose Range Complex
MRB	Motorized Rifle Battalion
MRR	Motorized Rifle Regiment
MTP	Mission Training Plan
MTS	Moving Target Simulator
MWST	Mounted Warfare Simulation Trainers
NBC	Nuclear, Biological, and Chemical
NCO	Noncommissioned Officer
NCS	Net Control Station
NGB	National Guard Bureau
NTC	National Training Center
O/C-T	Observer Controller/Trainer
ODS	Operation Desert Storm
OPFOR	Opposing Force
OPORD	Operations Order
ORE	Operational Readiness Evaluation
POM	Preparation for Overseas Movement
RC	Reserve Components

RG	Readiness Group
RTB	Regional Training Brigade
RTC	Reserve Training Concept
RTD	Resident Training Detachment
RTT	Reserve Training Team
STX	Situational Training Exercise
TCE	Tank Crew Examiner
TCPC	Tank Crew Proficiency Course
TDA	Table of Distribution and Allowances
TGST	Tank Gunnery Skills Test
TOE	Table of Organization and Equipment
TRADOC	U.S. Army Training and Doctrine Command
USAR	United States Army Reserve
XO	Executive Officer

Chapter One

INTRODUCTION

BACKGROUND

Three Army National Guard (ARNG) combat brigades were called to active duty during the preparations for the Persian Gulf War. It took considerably longer to prepare these brigades for deployment than many had anticipated, given that they were regarded as the most ready combat units in the Guard. Both Congress and the Army were concerned about the adequacy of peacetime training programs to prepare these brigades for deployment. As a result, in the summer of 1990 U.S. Army Forces Command (FORSCOM) implemented a program called Bold Shift to test a set of actions designed to improve the readiness of the Reserve Components (RC). The Arroyo Center assessed the status of units in the Bold Shift program at the end of 1992.[1] Also in 1992, Congress directed the Army to institute a program to increase active-duty support for peacetime training of ARNG combat and early-deploying units.[2] Subsequently, the Department of Defense's Bottom-Up Review laid out a plan for 15 "enhanced readiness" brigades, which are expected to get high priority for such support and other scarce resources.

Although the Army has taken steps to improve premobilization training, postmobilization training remains an issue. During the preparations for the offensive phase of the Gulf War (Operation Desert Storm or ODS), active-duty combat units were available to assist with the training of three ARNG brigades mobilized for that conflict, but significant force reductions have occurred since then. As a result, active units are not likely to be available to assist with the postmobilization training of the enhanced brigades. Under any foreseeable set of strategic conditions in which ARNG brigades deploy to an active combat theater, the United States will have first deployed available active component (AC)

---

[1]Sortor et al. (1994).

[2]This direction is contained in the so-called Title VII and XI legislation included in the FY92 and FY93 defense authorization acts. Broadly described, they direct dedicated active component support to peacetime reserve training along with other readiness enhancements.

heavy combat units from CONUS. The active units would be preoccupied with their own predeployment activities and would be unavailable to assist the ARNG units as they did during ODS.

Furthermore, many support units would not be available to help prepare the ARNG combat brigades. Almost all AC and high-priority RC support units, known as Contingency Force Pool (CFP) units, would be mobilizing and deploying to support AC combat forces. Under most circumstances, these units would have priority for resources over ARNG brigades. This means that neither the AC nor high-priority RC support units can be expected to be available to support postmobilization training of the ARNG enhanced brigades.

The question arises, What resources would be needed to support such training, and where would they come from? Needed resources include

- training sites
- training personnel
- opposing force (OPFOR), and
- installation support.

The Arroyo Center has developed a model to identify the resources necessary to carry out the postmobilization training of enhanced ARNG armor or mechanized infantry brigades.

We note that even as we write this document, the Army is in a state of enormous flux. Many of the assumptions we have made and much of the data we have drawn upon to construct our model could change significantly as a result of the turmoil in the Army at large. Still, we believe that although the details may vary, the methodology used will remain valid for bounding the level of resources needed to support reserve training after mobilization.

PURPOSE

This report has five purposes.

- It describes the training model. We characterize it as a model rather than a training plan because its purpose is to provide the basis for assessing the types and quantity of training resources needed to carry out each training event reflected in the model.
- It quantifies the resources needed to support the model, identifies sources for those resources as reflected in current Army plans, and identifies resource constraints.

- It proposes some alternatives for executing the training model.

- It illustrates the rate at which the various alternatives could generate trained brigades.

- It discusses the implications of this analysis for a broader set of policy and resource issues.

RESEARCH APPROACH

Figure 1.1 depicts the approach we followed.

To determine the resources required to train an ARNG brigade, we first reviewed the characteristics of the units to be trained, in this case seven enhanced ARNG brigades. We then surveyed the resources available and made a number of assumptions necessary to develop a training model. For example, we made assumptions about the types of missions the ARNG brigades would have to perform in theater.

We then developed a detailed training model, outlining a postmobilization schedule of training events for each element in the brigade. We structured the model to meet the most demanding mission, a brigade fully prepared to per-

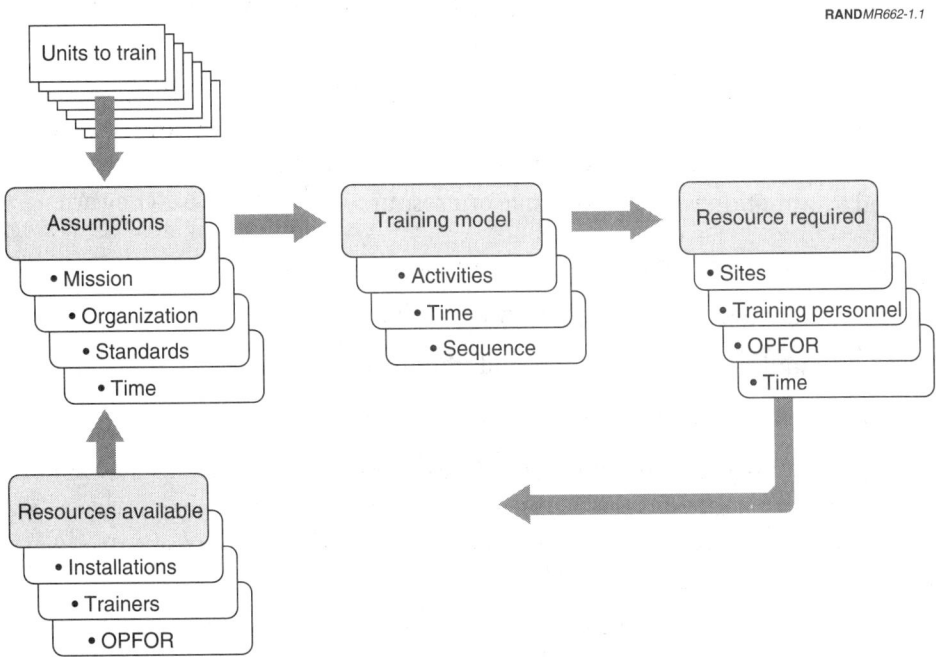

Figure 1.1—Research Approach

form complex tactical missions in a combat theater and deploying as soon as possible after mobilization. We also sequenced the training events so that earlier stages laid a foundation for later ones.

We then calculated the resources needed to execute each event in the model and totaled these resources by category (trainers, training support personnel, OPFOR, ranges, and maneuver areas) and type (e.g., the grades and specialties required for trainers). This process allowed us to quantify the resources required to train a single heavy brigade including its support elements.

We next examined different alternatives for implementing the model, that is, different numbers and types of sites. The resources required for the most feasible alternatives were compared with available resources to determine the number of brigades that could be trained simultaneously and the rate at which brigades could be trained for each alternative.

HOW THIS DOCUMENT IS ORGANIZED

The remainder of this document is organized into six chapters. The next chapter describes the training model, and Chapters Three, Four, and Five describe the resources required by the model and the resources available following mobilization to support the training of the ARNG brigades. They address, respectively, training sites, personnel, and opposing forces. Chapter Six describes alternatives for implementing the model and the resources required; it also describes the force generation that could reasonably occur for each. The final chapter presents observations and conclusions.

The document also contains six appendixes. In general, these contain the detailed information we used to derive our resource requirements. They are:

- The training model
- Site maneuver area and gunnery capabilities
- The requirements for trainers and training managers
- The sources of trainers and training managers
- The methodology for determining garrison support
- OPFOR requirements

Chapter Two

DEVELOPING THE TRAINING MODEL

We designed the model as a tool to help quantify the resources needed to support postmobilization training of an enhanced ARNG armor or mechanized infantry brigade. To make the tool functional, we had to develop the model in sufficient detail so that we could determine the full complement of resources. We considered maneuver area, ranges, equipment, people to support the training (both directly, such as observer/controllers, and indirectly, such as maintenance and range support personnel), and support required from the installation.

Several goals shaped the construction of the model. First, we wanted to ensure that it produced a brigade trained well enough to carry out the most demanding mission: entering combat shortly after arriving in theater. Also, we consciously tried to keep the time required to a minimum. Thus, we employ parallel training as much as possible, and we provide sufficient assets so that brigades can focus on activities that directly contribute to their ability to perform in the combat theater. The model calls for outside trainers and support personnel to perform any activities that do not contribute to this objective. However, we also wanted to ensure a feasible model, so we included time for maintenance, training preparation, and retraining.

We emphasize that this model embodies RAND's estimates; it is not official DoD policy. DoD statements have expressed the goal of having the brigades ready to deploy in 90 days.[1] Our training model requires slightly more time, a total of 102 days, based on detailed analysis of the steps needed to prepare brigades for combat missions. However, this difference is not significant for our main purpose, which is to identify the resources required for postmobilization training.

---

[1]See, for example, the report of the Bottom-Up Review (Aspin, 1993, p. 94) and subsequent DoD statements before Congress (Lee, 1995).

SOURCES AND REVIEW

We drew on a number of sources to develop the model. The basic training events were selected based on the gunnery and training doctrine as outlined in Army Training and Evaluation Plans (ARTEP), Mission Training Plans (MTP) and Drills, Field Manuals, and Department of the Army (DA) Pamphlet 350-38, *Standards in Weapons Training.* Other mobilization events were based on the requirements and guidance in FORSCOM mobilization documents.[2] We also used training programs performed by ARNG heavy brigades mobilized during Operation Desert Storm,[3] current AC postalert and RC postmobilization training plans,[4] current AC training programs used to prepare brigades for National Training Center (NTC) rotations, and the NTC rotational programs[5] themselves to determine the sequence and length of time devoted to each training event.

Once we had developed the training model, we circulated it widely for review and comment, in both written and briefing form. We provided written copies to members of the Forces Command (FORSCOM) and the Training and Doctrine Command (TRADOC) staff. We briefed the model to the National Guard Bureau (NGB), members of the FORSCOM and TRADOC staff, and the commander of the National Training Center and members of his staff. At the NTC, either the chief trainer or the operations officer of the specific observer/controller teams reviewed the model in detail and provided comments.

ASSUMPTIONS

Developing the model required us to make assumptions in different areas. These include the training level of the units and missions that they would be trained to perform, the training status of the unit when it arrives at the training site, the status of its personnel, equipment, and resources, and the planning and preparation that has occurred before the unit's arrival.

Training and Missions

The most influential assumption is that the brigade must be trained well enough to enter combat shortly after arriving in theater. Our review suggests

---

[2]These include ARTEP MTPs 71-3, 71-2, 71-1, 7-8, 17-237-10; FMs 17-12, 23-1, 25-100, 25-101; and DA Pamphlet 350-38. All Army training documents are listed under their own subheading in the Bibliography.

[3]Lippiatt, Polich, and Sortor (1992). See also NTC Operations Group (1990b), 4th Infantry Division (Mech) (1991), and 5th Infantry Division (Mech) (1991a).

[4]2nd Armor Division (1994), 4th Infantry Division (Mech) (1994), 1st Infantry Division (Mech) (1994), and 1st Cavalry Division (1994).

[5]NTC Operations Group (1994a).

that achieving this level of training requires substantial resources, including a competent opponent, outside expert trainers and training support personnel, and adequate facilities and support for training.

We assume the brigade has to be trained to execute three major missions at time of deployment: movement to contact, deliberate attack, and area defense. The model provides the time and resources for the brigade and its subordinate elements to be trained in all the supporting collective tasks and Battle Operating Systems functional tasks needed to execute these battlefield missions.

We derived our assumption about the missions primarily from the recommendations of the working group of the Enhanced Brigade Task Force.[6] We also had input from the Department of the Army staff. This complement of missions would seem to be the minimum required for a brigade-sized unit going to a combat theater. A unit so trained can defend itself, sustain itself, move tactically, and attack.

The training model described here is not intended as a template solution for all postmobilization training. Different assumptions about the postmobilization mission would lead to a different training model. For example, if the brigade were to replace an active unit stationed overseas that had deployed to the combat theater, much of the training in the current model could be deferred or eliminated.

Premobilization Readiness

We assume, first, that the brigade has trained in accordance with the Bold Shift strategy,[7] emphasizing platoon-level maneuver, gunnery, and command, control, and sustainment training programs for all elements of the brigade.[8] And we assume that the brigade training readiness at time of mobilization matches or exceeds that of the better brigades at the end of FY93.[9] Such a brigade would have had 70 percent of its tank or Bradley crews at the last Annual Training (AT)

---

[6]Headquarters Department of the Army (1994).

[7]U.S. Army Forces Command, *Regulation 350-2*, Fort McPherson, Georgia, June 1995.

[8]At the direction of Congress, the ARNG is testing several simulation training initiatives under a program called "Project SIMITAR." These appear promising and, if successful and implemented, may reduce the amount of postmobilization training required. However, we believe that all the events in the model—and thus the resources needed to implement it—will essentially remain unchanged. Most trainers we have interviewed support this conclusion.

[9]RAND conducted extensive visits to round-out and round-up brigades during Annual Training in 1992 and 1993 and collected data to make approximations on the level to which they have reached established readiness goals. For descriptions of the analysis results, see Sortor et al. (1994). Extensive data on personnel readiness and Annual Training attendance were also drawn from FORSCOM's Training Assessment Model (see U.S. Army Forces Command, *Training Assessment Model (TAM)*, U.S. Army Forces Command, Fort McPherson, Georgia, May 1993).

period, although some of those crews would have been composed of members of several crews. Most—85 to 90 percent—of the crews firing during AT would have qualified on Table VIII.[10] Platoon maneuver would have been accomplished, but for offensive and defensive lanes only, in daylight only, and for a subset of the required tasks.

A brigade at this level of proficiency includes crew members trained well enough to go to a tank or Bradley Table V range after three days of refresher training. The platoons are trained well enough to go through a full set of specialized platoon situational training exercises (STX) after four days of refresher training, the primary purpose of which is to integrate replacements.

These assumptions may be optimistic, because several factors combine to make it difficult to sustain collective skills throughout the year. Thirty percent do not participate in collective training with their unit during AT.[11] Companies typically experience between 25 and 32 percent attrition each year or between 2 and 3 percent each month.[12] Job turbulence is typically twice the rate of attrition, averaging between 50 to 60 percent annually, and crew turbulence approaches 60 percent annually.[13] Thus, only 90 days after Annual Training, a tank company could have lost 15 percent of its people, and of those remaining, 12 percent would have changed jobs and 12 percent would have changed crews. Furthermore, only 75 percent of the assigned personnel are qualified in their primary military occupational specialty (MOS).[14] Because collective skills are difficult to sustain throughout the year for the reasons cited, postmobilization training plans must provide an opportunity to stabilize crews and units within the brigade and also provide the appropriate level of collective refresher training.[15]

---

[10]"Tables," numbered from I to XII, are the Army's classification scheme for ranges where tanks or Bradley Fighting Vehicles engage different sets of targets under different conditions. The tables begin with single stationary vehicles shooting at stationary targets and progress to multiple vehicles engaging targets on the move.

[11]Most of these individuals were attending required schools or, because of scheduling conflicts, participated in Annual Training with another unit or at their home armory.

[12]Sortor et al. (1994) and U.S. Army Forces Command, *Training Assessment Model (TAM)*, U.S. Army Forces Command, Fort McPherson, Georgia, May 1993.

[13]Job turbulence is a change in an individual's primary job during the year. Crew turbulence refers to a change in the track commander-gunner combination.

[14]The NGB has initiated a readiness enhancement program to improve personnel readiness, among other areas. However, we have observed little change in this area. The reason may well be the current series of adjustments made to accommodate force reductions; these could preclude any significant near-term improvement.

[15]Active component units suffer similar attrition and turbulence, but have more opportunity during the year to sustain collective skills. Most AC combat units do, however, have postalert training plans that provide for collective training before deployment if time is available.

We presuppose, further, that the support units and command-and-control organizations have enough personnel and are trained well enough to support field training for the entire brigade while simultaneously conducting their own training.

Personnel Readiness

The model assumes that the readiness status of all units is at least C-1 for personnel; that is, 90 percent of required personnel are present, qualified, and stabilized by M+18 (that is, 18 days after the brigade is mobilized), approximately the date when gunnery and platoon-level training begins for all units. As noted above, we make several assumptions about the level of proficiency achieved at Annual Training.

Equipment Readiness

We assume that the unit has all of its major combat systems on hand, and, further, that almost all of that equipment is fully operational by M+18.[16]

We further assume that the supply and maintenance support systems are able to bring equipment on hand and equipment status ratings to C-1 by M+18 and that the training site will have sufficient maintenance support (facilities, personnel, parts) to begin training and to maintain the operating tempo (OPTEMPO).[17] This is true even though the brigades will mobilize concurrently with early-deploying AC and RC units (including AC divisions) and compete with theater requirements, both of which will have priority for resources.[18]

---

[16]Again, this may be an optimistic assumption. In our discussions with brigades, divisions, and various headquarters, we have heard that there are considerable backlogs for some maintenance areas. Although we have no detailed data to document the types of backlog that may present problems, any serious problem with equipment could extend the time needed to prepare for deployment.

[17]A 14-day alert period to improve readiness and to remedy shortfalls in personnel, equipment status, and equipment on hand may be required to accomplish this. If, on the other hand, units have already achieved high readiness, such an alert period might shorten the postmobilization, preparation for overseas movement (POM), and preparation for transportation requirements.

[18]Repair parts could be a critical resource constraint. Although examination of this area was beyond the scope of our study, some indications suggest that the availability of sufficient repair parts to simultaneously support maneuver forces in a combat theater, AC divisions preparing for deployment, and the OPTEMPO required to conduct training for up to six heavy enhanced brigades and three OPFOR brigades may well be a limitation. Recent reductions in maintenance funding for AC maintenance activities and possible maintenance backlogs in Army National Guard maintenance facilities both support a contention that availability of repair parts may be a significant constraint.

Transportation and Equipment Sets

The model further assumes that transportation will be available for movement to the training site.

Resources

Trainers and OPFOR. A key assumption is that no AC TOE units are available to support mobilization. Sufficient trainers, training support personnel, and OPFOR will be ready, and the organizations responsible for these functions will be prepared to start training at M+18

Collective training site capabilities. In our model, the enhanced brigade mobilizes at one station and trains at a second site. Most mobilization stations cannot support brigade-level training. In cases where the mobilization station and the training site are the same, time can be reduced.

The model assumes the training site has the following capabilities and facilities:

- **Two maneuver areas available.** One area is needed for maneuver training up to brigade-level deliberate attack or battalion defense in sector, a second area for battalion deliberate attack or movement to contact. The two areas must be available for simultaneous operations prior to mobilization.

- **Adequate gunnery ranges for companies/cavalry troops.** These ranges should permit Bradley Fighting Vehicle/Tank Gunnery tables to be conducted sequentially, with each company/troop separately firing a Table IV, V, VI, VII, VIII, and XII at one time. (Generally this requires a Multipurpose Range Complex (Table XII), Multipurpose Target Range (Table VIII), and ranges for Tables IV, V, VI, and VII and Dismounted Infantry Live Fire. Some but not all major CONUS maneuver installations can meet this requirement.)

- **Company-level Combined Arms Live-Fire Exercise (CALFEX) range.** The model requires a CALFEX range that can be operated while allowing concurrent battalion task force–level maneuver training. The range must be able to support two company CALFEXs at one time.

- **Simulation support.** Adequate tank and Bradley Fighting Vehicle Conduct of Fire Trainers (COFTs), Stinger training facility (Moving Target Simulator or MTS), MILES sets, and command-and-control training simulations (JANUS or BBS) must be available at the training site.

Training methods. The model was based on several assumptions about the methods that would be used to conduct the training. These are:

- **Maximize parallel activities.** All brigade organizations train concurrently. This requires sufficient trainers and training managers, OPFOR, and command-and-control structure so that the brigade is free to focus on its own preparation for deployment. For example, it assumes that ranges are "roll-on/roll-off" for the unit; that is, none of its personnel are required to fulfill range support functions.

- **Follow the reserve training concept.** That is, training is progressive and begins with leader, individual, and small unit preliminary training and rehearsals preceding collective evaluation. The manner of training will enhance unit integrity. Collective maneuver and gunnery training will be conducted, supported, and controlled by the RC chain of command from field locations. RC support organizations will perform sustainment activities.

- **Conduct C2 training in parallel.** Higher-echelon units—brigade and battalion commanders and staffs—whose primary mission is command and control or sustainment will train on those functions at the same time the units they support are going through structured training. Time is allocated for both types of training.

- **Allow time for retraining and maintenance.** The model includes such time, but it is held to a minimum. Approximately every third day, the schedule provides time for either maintenance, retraining, or both conducted concurrently. Given the fact that the training period is considered intense (75 continuous days), this apportionment is optimistic. Adequate premobilization training, organizational management, physical fitness and equipment readiness levels, and fully effective use of time will be required to maintain this schedule.

Planning and Preparation for Postmobilization Training

The general scenario we envision is that a brigade-sized unit arrives at an Army training site twelve days after call-up and that it begins training six days after that. The people supporting the training would have to accomplish a myriad of tasks before that unit could begin training. The area for each training event would have to be identified, trainer and OPFOR for each training event identified, physical reconnaissances and preparation activities at each site accomplished, ammunition ordered, range support personnel organized, and so forth. We assume that all the necessary planning, preparation, and coordination has occurred by the first training day. In our research for this project, we found no installations or AC training organizations with detailed plans for postmobilization training or executing preparation activities, although all agreed that these would be needed to execute a complete postmobilization training program in a timely manner. The basic issue is availability and allocation of resources to perform these planning functions in peacetime. Given peacetime demands in an Army that is downsizing, it is in many ways a more difficult problem for the active component to allocate people for such planning than to identify resources needed to support postmobilization activities.[19]

Character of Assumptions

In building the model, we have attempted to use realistic assumptions. And we have built in sufficient flexibility so that everything does not have to go exactly as planned. That said, the overall set of assumptions tends to be optimistic. Should a significant problem occur, say a company fails to qualify on the higher-level tables or task force maneuvers prove unsatisfactory, either the schedule would have to slip or units would have to deploy without having demonstrated proficiency in some areas.

THE TRAINING MODEL[20]

In the model, the entire postmobilization period takes approximately 102 days, which divides into five general categories. Table 2.1 lists these categories and approximate time allocations for battalion maneuver units in days for each.

Unit activities are identical across unit types in four of the five categories. Considerable variation by unit type occurs in the individual, squad, platoon, and gunnery category.

We discuss each of the categories below. We begin with a general description of the initial phases of mobilization, and continue with a discussion of the training activities of the two major components of the brigade: the maneuver elements (armor or mechanized infantry battalions and the armored cavalry troop) and the brigade "slice" (supporting field artillery, combat engineer, and forward support battalions).

---

[19]This was true during our visits in the summer of 1995. Of the six installations we examined as potentially adequate for postmobilization training, only Fort Lewis/Yakima and Gowen Field were formally tasked to be postmobilization training sites for an enhanced brigade, and this was because they were already a mobilization station for a brigade. None of these sites had a detailed plan for postmobilization training. However, this must be put into context. At that time only one of the reserve training brigades (RTBs) (the organization that would be responsible for conducting training for the enhanced brigades under CONUSA) had been established, and it had been in existence for less than a year. Moreover, the Army was transitioning from four CONUSAs to only two, which would assume responsibility after transition from the round-out/round-out concept to the enhanced brigade concept in FY96. The basic issue is that premobilization planning and preparation for postmobilization training will require additional resources.

[20]Appendix A contains a more detailed allocation of the time allowed for the different training events. It also provides a summary description of the different events.

Table 2.1

Mobilization Categories

	Days
Initial preparation and movement to training site	17
Individual, squad, platoon, and gunnery training	33
Task force organization and company training	17
Battalion task force– and brigade-level training	25
Maintenance, equipment services, final preparation	10
Total	102

NOTE: The time allocations are approximate because they vary by type of unit. For example, four days are allocated for platoon drills for armored units, and five for mechanized infantry units.

This model posits a brigade with the following structure:

- Brigade headquarters
- 1 tank battalion
- 2 mechanized infantry battalions
- 1 field artillery battalion
- 1 combat engineer battalion
- 1 forward support battalion
- 1 armored cavalry troop
- 1 air defense artillery battery

Initial Preparation (Days 1–17)

The first 17 days of the training model are identical for all units. Twelve days are allocated for movement to the mobilization station, preparation for overseas movement (POM) activities, preparation of equipment, and movement of personnel and equipment to the collective training site. This period is somewhat shorter than the time required during ODS, but it seems reasonable given increased readiness in these areas reported by Operational Readiness Evaluation (ORE) teams.[21]

---

[21]Sortor et al. (1994).

The first five days at the training site are dedicated to receiving and integrating replacements and finalizing maintenance actions to bring equipment up to mission capable (MC) standards required to sustain the intensive maneuver and gunnery training portion of the model. The current personnel and equipment readiness status of the brigades makes this a somewhat optimistic estimate. Concurrent with these activities, individual training on survival and other critical individual tasks and crew COFT training are conducted to expedite transition to collective maneuver and gunnery training.

Minimum days are allocated to maintenance to bring equipment up to MC standards prior to start of training. Adequate premobilization equipment readiness and training of maintenance personnel and availability of maintenance support will be required for this schedule to succeed.[22]

Individual, Squad, Platoon, and Gunnery Training (Day 18–50)

During this period, gunnery, individual, squad, section, and platoon-level training takes place for all elements of the brigade. Additionally, structured command-and-control training is conducted for battalion and brigade commanders and staff, and the support elements of the brigade (field artillery battalion, engineer battalion, forward support battalion, and air defense artillery battery) complete their internal training. The objective is for command, control, and support elements to be prepared to participate in the company-level training beginning on Day 54.

Maneuver platoons. Gunnery training begins with a tank or Bradley gunnery skills test and tank or Bradley Crew Proficiency Course (TCPC and BCPC). Companies go through Tables V–XII (including TOW tables for Bradleys). In Bradley infantry battalions, dismounted infantry individual and crew-served weapons training and live-fire squad defensive and offensive exercises are accomplished concurrently with tank and Bradley crew gunnery tables to prepare both elements for qualification on Table XII. An average of 18 days is allocated for each battalion to accomplish gunnery and squad live-fire training. Two more days are provided for individual and crew-served weapons qualification. This schedule is somewhat faster than ODS experience, but the presence of "turnkey" ranges and improved premobilization training are assumed to make this schedule possible.

Maneuver training starts with a MILES training day, followed by four days of refresher platoon- and squad-level battle drill and movement training. This

---

[22]Some observers believe this to be optimistic given a perceived maintenance backlog at Army National Guard maintenance facilities and the maintenance status of equipment drawn during AT.

training takes place without OPFOR. The purpose is to integrate new personnel and prepare for force-on-force training.

Maneuver training continues with platoon STXs. Tank and Bradley Fighting Vehicle (BFV) platoons execute four lanes, and scout platoons execute five. These STXs include all the tasks needed for platoons to enter higher-level collective training. Nine days are allocated to this training, including practice, retraining, and maintenance time. The total of fourteen days allocated to platoon drills, lanes, and STXs takes somewhat longer than the average during ODS. But this additional time for platoon drills does not expand the total time. Because of the gunnery range constraints and the parallelism of the training model, shortening the time for platoon training would not reduce the time required for the overall schedule.

Figure 2.1 depicts the major training events for the armor and two infantry battalions and the scout platoons that take place within the first 51 days. The parallel training depicted here, in which all units are training simultaneously but on different events, differs from the sequential, building-block training used in preparing the RC brigades for ODS.[23] Our assumptions call for an installa-

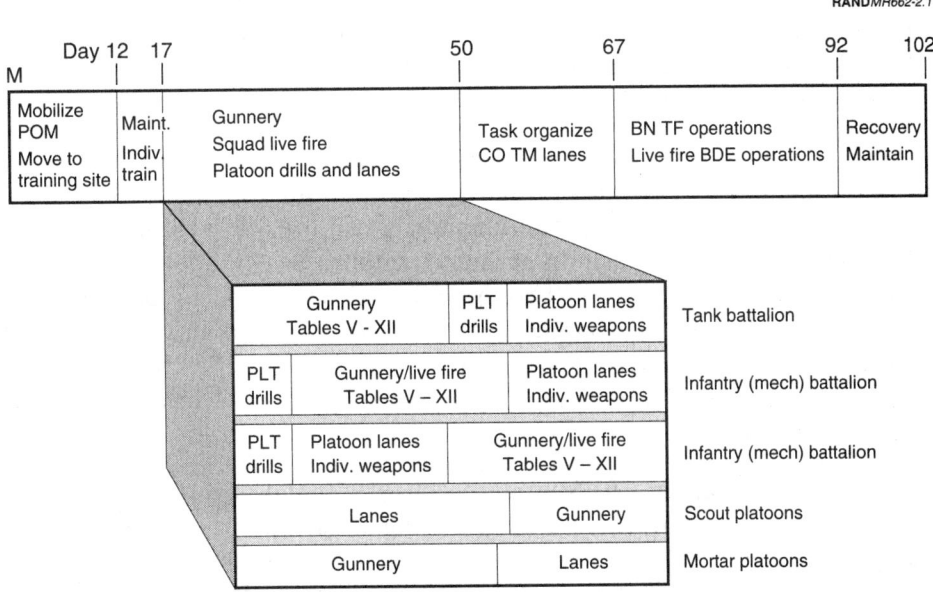

Figure 2.1—Brigade Maneuver Battalion Training Model Through Day 50

---

[23]The Armor School at Fort Knox is exploring alternative gunnery qualification strategies to be tested in the ARNG. If these strategies are accepted and require less time, the training model depicted here could be shortened accordingly.

tion with one set of gunnery ranges. The ranges become the binding constraint, and as Figure 2.1 indicates, other training events are scheduled around the time a unit is firing the gunnery tables.[24] This training model assumes that by Day 54 all units are ready to participate in company team–level STX.

Maneuver battalion support elements. Training for maneuver battalions' mortar and scout platoons is centralized at brigade level and follows a similar progression to the tank and BFV platoons, including both the maneuver and live-fire portions.

Battalion task force training for combat service support (CSS) elements initially involves individual- and section-level training on technical functions. Collective training involves a combination of functional support for task force elements and structured functional STX (for example, casualty evacuation and treatment). From Day 18 through completion of brigade operations, sustainment functions are performed from a tactical configuration with "cover-down" assessment and coaching.

Preliminary command and staff training. Command and staff training for brigade and all battalion/company commanders and staffs is conducted in parallel with crew-, platoon-, and company-level collective training. This training lasts 39 days and begins with individual and section-level technical and functional training, progressing from basic to complex leader-training exercises supported by simulations.

Command and staff training is structured to allow company commanders to participate in platoon STX and to allow battalion commanders and their staff to participate in company STX. Approximately half the day is devoted to structured C2 training. Throughout gunnery, platoon, and company STX training phases, commanders and staff command and control their subordinate elements, learning their capabilities in the process. During periods when the senior staff are involved with subordinate unit training, C2 training for the command-and-control units continues at the appropriate level using junior staff.

Company commander, fire support team (FIST), and CSS training are conducted before company STX training. During initial mobilization activities and gunnery training, individual doctrine and functional training are conducted for these sections to prepare them for collective training at the company-team level. During platoon STXs, the company commander, executive officer, and FIST perform their normal C2, fire support, and CSS functions. The training ex-

---

[24]In the detailed model, the first company actually enters gunnery training on Day 15, with following companies starting at two-day intervals. This provides time for all but the first company to accomplish any required maintenance, train individuals, or train crews on the COFT as needed.

perts also mentor the company commanders during the platoon exercises (the same mentoring also occurs for battalion commanders and staffs during the company exercises).

Training for other brigade elements. Training for the field artillery, air defense artillery, combat engineer, and forward support battalion parallels maneuver battalion training for the first 18 days. After Day 53, from the company team STX portion through the completion of brigade operations, these elements support and participate in maneuver company, battalion, and brigade training events. The commanders and staffs of these organizations participate in the preliminary C2 training described previously. Figure 2.2 compares the training of the brigade slice elements with the maneuver units. The numbers at the top of the second bar of the figure reflect cumulative days, and those at the bottom the time required by each segment. The top bar illustrates the C2 training for the higher echelons.

The field artillery battalion has programs for the firing batteries, company FIST, fire support officers (FSOs), fire support sections (FSSs), and combat service support elements. Firing battery training proceeds according to the new artillery tables, which begin with crew and fire direction center (FDC) refresher training and certification and move through battalion live fire. This training culminates with a four-day battalion external evaluation. This program takes 42 days and is longer than ODS experience. Time is not an issue because this schedule allows the batteries to complete training and participate in maneuver battalion and brigade collective training.

FIST training takes place on several concurrent events and requires careful scheduling. FISTs participate in maneuver platoon STX and artillery and mortar live-fire exercises, and they receive structured functional training. This functional training includes use of the Forward Observer Training System. Beginning with the company team STX, FISTs join and participate in maneuver training. FSO/FSS elements participate in battalion and brigade C2 training.

Field artillery CSS elements' training parallels that of maneuver battalion CSS elements, individual and section functional training to start, followed by functional support from a tactical configuration concurrent with special STX.

Engineer battalion training parallels maneuver battalion training and includes MILES, battle drills/critical tasks, gunnery, and platoon and company STXs. There is no battalion-level field training exercise, but the engineer battalion does have battalion-level command post exercise (CPX) and command field exercise (CFX) training. Engineers participate in company training and maneuver battalion and brigade field training. The engineer battalion headquarters participates in brigade C2 training, and its companies play a part in maneuver battalion C2 training. Platoon STXs last twelve days, with eight separate training

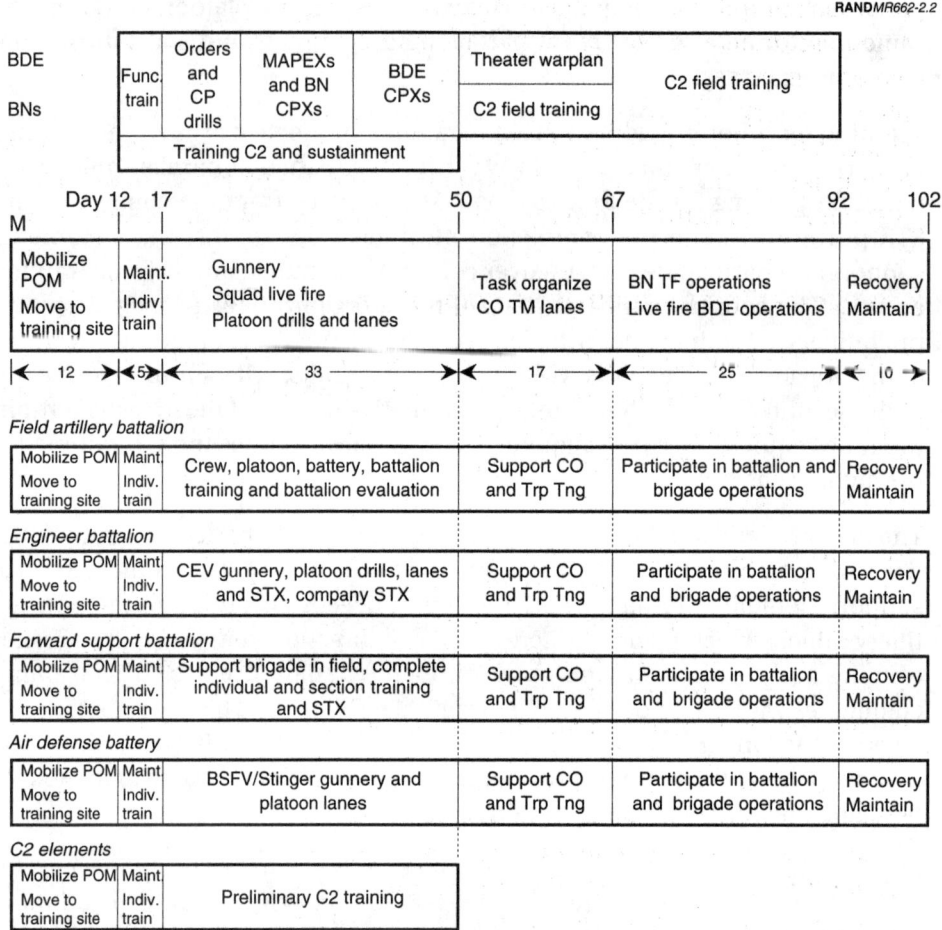

Figure 2.2—Brigade Slice Training

events selected to train on the tasks needed to support brigade movement to contact, deliberate attack, and defense. Company STXs last seven days and contain four training events. Combat Engineer Vehicles (CEV) gunnery and demolition qualification lasts four days.

Forward support battalion training consists of individual and section functional training, functional STX, and support of other brigade elements from a tactical configuration.

Air defense artillery battery training centers on gunnery training for BFV (BFVs in the two BSFV platoons are integrated into Bradley gunnery training with the two mechanized infantry battalions) and Stinger teams. BSFV Tables IX A and X A are also integrated with mechanized infantry gunnery. Stinger gunner evaluations are conducted at a Moving Target Simulator (MTS) simultaneously with BSFV gunnery. Stinger gunner training is completed in time to allow Stinger gunners to participate in Tables IX A and X A. BSFV Stinger platoon STXs are conducted for the remainder of this period, and the battery commander and platoon leaders participate in preliminary command-and-control exercises.

Task Force Organization and Company Training (Day 51–67)

Cross attachment takes place at Day 51. The first three days are allocated to reorganization and conduct of simple company team–level movement and sustainment exercises without OPFOR. Company team–level operation procedures are practiced at a basic level to prepare for the more demanding force-on-force lanes.

The parallel training and the need to move to company team training at Day 54 leave little margin for error. A unit having a major problem during the tank-table firing could not spend more time on the range to resolve it, since another battalion would require the range. The only options would be to slip the schedule for the brigade or to proceed, hoping to rectify the deficiency during the final deployment preparation or after arrival in theater.

Six STXs are conducted for each company team, with 14 days allocated to this training. This allocation is less than the ODS average. Because of the more difficult nature of the company exercise, a preparation day precedes each STX. (Platoon-level STX also included preparation, but it occurred on the same day as the exercise.) As with the platoon lanes, task force, C2, and CSS elements perform their normal functions.

Not all company teams perform the same exercises. For example, infantry companies and some infantry-heavy company teams have assault STX, while tank-heavy company teams and other infantry-heavy teams have breach STX. The tasks included in the company team STX are selected and lanes designed to prepare for battalion task force–level training in movement to contact, deliberate attack, and defense in sector. The scheme provides an opportunity for alternate METT-T specific tasks, e.g., attack in an urban environment. So the STX lanes could differ, but the resources required would be about the same.

The armored cavalry troop and the air defense artillery battery, with its Stinger platoons, perform specialized company-level STX during this period. The armored cavalry troop lanes focus on reconnaissance and security missions, and the air defense artillery battery lanes focus on area defense missions executed by Stinger platoons.

Battalion Task Force– and Brigade-Level Training (Day 68–92)

Battalion and brigade-level training and company team CALFEX training is conducted over a 14-day period.[25] In the first 10 days, each battalion task force conducts two deliberate attacks, two defenses, and one movement to contact FTX. Three days are allocated for each battalion task force FTX to allow for a crawl-walk-run approach with full preparation and training. Time is available to conduct a CFX or full-scale rehearsal on the second day. There is also time to repeat the FTX if performance warrants additional training.[26]

This approach allows the brigade to prepare the order and control execution of each battalion task force operation and for brigade combat support (CS) and CSS elements to provide normal support. The last three days are a brigade-level FTX with all brigade elements.

Final Preparations (Day 93–102)

Beginning on Day 93, the brigade begins final preparations for deployment. This process lasts 10 days and includes maintenance required after 80 days of intense field training, final preparation for overseas movement, and loading activities.[27] Required periodic services will be performed as needed before deployment, and for many crews this is considered a structured training event.[28] A 10-day period is optimistic in that most crews and many maintenance personnel have never done these services before, and these activities are as much training as they are preparation.[29] They will need to know how to perform these tasks in theater.

SUMMARY OF THE TRAINING MODEL

The training model provides a tool for calculating the resources required to train one ARNG heavy enhanced brigade following mobilization. In designing

---

[25]Note that if the training takes place at Fort Irwin, a battalion task force CALFEX is possible. The other posts we examined would allow only a company team–level CALFEX.

[26]In these respects, this training differs considerably from a current NTC rotation, which is generally conducted at a "run" level throughout.

[27]Active component units require 5 to 6 days to recover and prepare equipment for turn-in at the National Training Center after only 14 days of field training. These units get substantial help from contractor personnel.

[28]Most ARNG crew and battalion maintenance personnel do not conduct required periodic services in peacetime. These are typically accomplished by full-time Army National Guard maintenance personnel. For many, this will be the first time they have performed periodic services, and this preparation is a necessary part for deployment to combat.

[29]Most active battalions we examined required at least 10 workdays to accomplish periodic services.

that model, we made a series of assumptions, the most influential of which are that the brigade needs to be trained well enough to enter combat shortly after arriving in theater, that no AC combat or support units would be available to assist during the postmobilization training, and that the training should occur as quickly as possible. The assumptions tend to be optimistic, although we believe they are prudent. However, should they not hold, the quality of the training or the time it takes (or both) will suffer.

The model results in a training period of slightly over 100 days. We emphasize that this model is not held up as a template solution for the postmobilization training of ARNG heavy brigades.

Chapter Three

SITES FOR IMPLEMENTING THE MODEL

Having discussed the model, we turn to the resources required to execute it. This chapter describes the locations that could serve as a brigade-level training site. It outlines the assessment criteria used and discusses the installations that meet the criteria.

CRITERIA FOR SITES

We evaluated the installations with respect to five sets of criteria:

- Gunnery ranges

- Combined arms live-fire exercise (CALFEX) capability

- Maneuver space

- Availability of facilities

- Installation facilities support

To execute the model in the time allotted, the installation must have sufficient maneuver, CALFEX, and gunner capability to allow different types of training events to occur concurrently. The most significant effect results from the requirement for two company-team CALFEXs and two battalion FTXs simultaneously. The installation must also have enough maneuver room for two brigades to maneuver at the same time. This requirement imposes a larger demand for maneuver area than many active component installations experience during peacetime training, when the density of training events and units training concurrently can be reduced and still meet peacetime training requirements.

Gunnery Ranges

For an installation to serve as a postmobilization training site, it must have the ranges to conduct five categories of gunnery exercises: Tables V–XII, Bradley and Tank Proficiency courses, TOW qualification, squad live-fire exercises, and mortar and field artillery live-fire exercises. In addition, the facilities have to enable the events to be executed to standard, i.e., as prescribed by the field manuals in terms of target types, engagement ranges, and so forth. The ranges should be automated to facilitate meeting the model's timelines. Finally, the ranges must be available at M+17. That is, they cannot be needed to support the training of a unit with higher deployment priority.

Combined Arms Live-Fire Exercise (CALFEX)

The installation has to have sufficient terrain to conduct two company-level CALFEXs, with offensive and defensive scenarios occurring simultaneously. The terrain must allow for tactically realistic scenarios and must allow both direct and indirect fires.

Maneuver Space

We drew general maneuver guidelines from TC 25-1.[1] The amount of maneuver space needed was determined by the largest requirement for maneuver areas on each day based on the number of events being trained, retrained, or prepared for training. Two events defined the largest maneuver area requirements. They were:

- Two simultaneous battalion task force–level force-on-force exercises (defense in sector and movement to contact)

- 10 simultaneous company team/cavalry troop lanes

The requirement for two battalion task force exercises translates into a requirement for two training areas, one measuring approximately 8×23 km and the other 10×30 km. The company team/cavalry troop lane requirements were up to 5×15 km each, with a more general average of 4×7 km. A smaller set of maneuver areas to support platoon lanes must also be available at M+17. The full complement must be available by M+51.

---

[1]We did not apply the Army's doctrinal space requirements (from Training Circular 25-1) literally to assess the maneuver area requirements. These requirements were modified based on the training objectives for each exercise or lane. For example, the company team maneuver area requirements were generally smaller than the TC 25-1 requirements to align with the smaller objectives for focused lanes training, compared to the broader training objectives for the company FTX requirements outlined in the TC. When we looked at each installation, we considered the actual maneuver corridors that would be used for each training event and the need to eliminate conflicting events.

Availability

Availability means that the ranges, maneuver areas, and facilities are available for the ARNG brigades. The most important consideration is the presence of AC combat divisions that would be carrying out their own final preparations and training for deployment and would presumably have priority for ranges, maneuver areas, and facilities. The most acute problem is at posts with later-deploying AC divisions. We also considered weather and environmental considerations that would limit training.

Installation Facilities Support

A final criterion for the evaluation of installations is the adequacy of garrison facilities. We include in this category maintenance, supply and medical facilities, railheads, and barracks. A lack of these facilities would not preclude training, but in many cases it would require TOE units to establish field facilities or depend on another Army post to provide some support functions (e.g., Fort Lewis providing support to Yakima). Additionally, lack of facilities for the personnel and organizations supporting training could potentially impede support and thus delay training.

Availability of MILES and support for it are also considerations. Somewhat more than two full brigade sets are required (one for the OPFOR and one for the ARNG brigade). Some backup sets are needed, as well as repair parts and a MILES maintenance system. This requirement exceeds what is available at any site except Fort Irwin, which means that any other site will require additional MILES equipment and support.

We also included the availability of facilities to support C2 (BBS and JANUS) and gunnery (COFT and MTS) simulation training. Again, a lack of these facilities would not preclude effective training, but it could either delay completion by adding a requirement to move brigade members to the closest available simulation sites or weaken the training by dispensing with the simulation-supported training.

A final consideration is the availability of supply items. These can range from individual items such as kevlar helmets to major repair part items such as a tank engine. Lack of repair items could seriously impede execution of the model, as it would directly affect execution of field maneuver and gunnery exercises.

After considering the difficulty of establishing adequate installation facilities at a location where they did not exist and the requirement for installation support personnel (discussed in the next chapter), we concluded that an AC installation would be required for the first set of heavy brigades to meet the timelines in the model. While an RC installation could be augmented, it would most likely take more than 18 days, which would delay the commencement of training and, in turn, deployment dates.

INSTALLATIONS THAT MEET THE CRITERIA

We evaluated installations across FORSCOM and TRADOC.[2] Our analysis shows that five installations—Forts Hood, Bliss, Carson, and Irwin, and the Yakima training area—meet the criteria.[3] FORSCOM also asked us to consider the Orchard Training Area, a reserve training site near Gowen Field in Idaho.[4] Although five installations meet the criteria, none is perfect. Figure 3.1 summarizes installation capabilities in each of the five areas. A white cell indicates the installation satisfies the particular set of criteria, gray that it has a significant limiting factor, and black that it cannot meet the criteria. A minus sign in a cell indicates a minor limitation. Appendix B contains a more detailed discussion of the evaluations.

Fort Hood has the best gunnery ranges of any of the installations studied. It also has excellent garrison facilities. A limitation is somewhat constrained maneuver areas that are subject to "light contamination" from nearby highways and civilian and post facilities. Availability could be delayed by the use of training areas and ranges by the one late-deploying division stationed there.

Fort Bliss has fully adequate maneuver areas. Its gunnery ranges allow all tables to be executed to standard, but the presence of only one automated range could limit throughput without effective range organizations and full gunnery support. Although additional targets would need to be provided, the CALFEX capability is adequate. Garrison facilities are sufficient, but the capacity to support two maneuver brigades is less than that of Forts Irwin and Hood. The availability of ranges and training areas will probably be good because the one early-deploying armored cavalry regiment stationed there now is scheduled to relocate to Fort Carson.

---

[2]For the sources we consulted, see the subsection "Maps" in the Bibliography. See also Headquarters Department of the Army (1993).

[3]As we conducted this study, many people expressed surprise that Forts Riley and Stewart did not meet the list of criteria. Neither installation meets our criteria for maneuver space for armored vehicles (two areas measuring approximately 6×23 km). Fort Riley simply does not have them (see Appendix B), and Fort Stewart, although it has large acreage, is broken up by heavily forested terrain or swamps.

[4]A large portion of this area is land that belongs to the Bureau of Land Management and is leased to the National Guard. Current legislation governing this leasing arrangement may preclude using the land for the type of training described in the model, but in a time of national emergency, enabling legislation could probably be obtained.

Fort Carson has adequate maneuver areas at Piñon Canyon; however, the distance between Piñon Canyon and the gunnery and CALFEX complexes and garrison facilities at Fort Carson proper could cause delay, especially without adequate transportation support. Also, weather could limit training during the winter months. Gunnery throughput could be limited by the fact that some of the gunnery tables must be done on nonautomated ranges, and gunnery standards are affected by the lack of required moving targets on some ranges. A significant limitation is that only one CALFEX exercise can be conducted at one time, which could add four to five days to the model. Availability of ranges and maneuver areas could be limited by the presence of a late-deploying divisional brigade.

Fort Irwin has excellent maneuver areas and a battalion task force–level CALFEX capability. Its gunnery ranges are not automated and do not have all the required moving targets on some of the preliminary ranges or for Table XII. Its garrison facilities are austere but probably adequate, with the lack of a Moving Target Simulator for Stinger training being the major shortfall. Its availability is good because no deploying AC combat units are stationed there.

Yakima, in Washington, is a possible brigade training site but has limitations, especially in terms of gunnery range capability. Currently the maneuver area is

RANDMR662-3.1

Criteria	Sites					
	Fort Hood	Fort Bliss	Fort Carson/ Piñon	Fort Irwin	Yakima	Orchard/ Gowen Field
C-day availability	~30	0	~30	0+	~30	0
Gunnery	+	Slow − XII nonstandard	Slow − Limited movers	Slow No movers XII nonstandard	−	
CALFEX		− −40 targets	Slow −40 targets	+		Space limited
Maneuver space	− Nonstandard BNTF			+ Instrumented		Nonstandard BNTF
Garrison facilities			Piñon austere	−	Austere	Limited

☐ Satisfies criteria   ☐ Has limitations   ■ Cannot meet the criteria

Figure 3.1—Installation Capability to Support Postmobilization Training

not sufficient for brigade and multiple battalion maneuver areas, but completing the acquisition of 500,000 additional acres in FY96 should remedy this issue. It has sufficient gunnery ranges to have concurrent Tables V–XII and squad live fire, but only the multipurpose range complex (MPRC) is currently automated and has moving targets. Plans call for adding another automated complex, which would reduce these problems. The two AC brigades at Fort Lewis are aligned with divisions in Korea and Hawaii, so availability could also be an issue. The installation support facilities are austere, and units training there would have to depend on Fort Lewis for much of their support.

Orchard Training Area (Gowen Field) has adequate gunnery capability and fairly good installation facilities. However, as we discuss in the next chapter, the absence of a full set of installation support organizations would impose an initial limitation. CALFEX capability and availability of adequate maneuver space are serious shortcomings. Its maneuver area is inadequate to support a company-team CALFEX. While additional federal lands in the area might reduce the problem, the maneuver area is still inadequate for battalion and brigade maneuver training. Additionally, there is some question about the availability of the additional federal land during postmobilization. The area is a protected habitat and operates under a provision that military use must not disturb that habitat. Expansion into additional areas or even a greater level of training tempo is currently not authorized, and early availability would need premobilization clearance. Additionally, as in Yakima, winter weather could seriously curtail training.

SUMMARY OF SITES FOR IMPLEMENTING THE MODEL

We examined six major CONUS installations against five sets of criteria we considered necessary for execution of the training model. The most demanding requirements in the five criteria sets are for two simultaneous company-team CALFEXs and two simultaneous battalion FTXs. While none of the sites fully met all criteria, five of the six examined were adequate. Because of possible delays in establishing installation support where it does not exist, active component installations are necessary for the initial brigades. Thus, adequate installations are available to train up to five brigades simultaneously on all events in the training model.

Chapter Four

TRAINING AND SUPPORT PERSONNEL

Having discussed the model and sites for its implementation, we now turn to the people required to support the training. These fall into two categories: personnel support for postmobilization training and installation support. The first category has two subcategories: (1) trainers, training managers, and training support personnel, and (2) OPFOR. Chapter Five discusses OPFOR requirements.

We identify two categories of training personnel: those who support training directly, such as the observer/controllers (O/Cs); and those who provide indirect support, such as running ranges, conducting supply and maintenance activities, replicating battlefield effects, and so forth.

For trainers and direct training support, this chapter describes our assumptions, what the support people do, the numbers needed, and the likely sources of personnel to fill the requirements. It also points out the comparative training strengths of the people from the different sources. These comparative strengths will weigh in the decision about options for implementing the model. Ideally, people would be employed in a postmobilization role that capitalizes on their comparative training advantages. Finally, the chapter identifies where sources fall short of requirements, both in numbers and skills.

Turning to installation support, we describe our methodology and the amount and type of installation support required. We also address the issue of the command and control of the postmobilization training sites.

PERSONNEL SUPPORT FOR POSTMOBILIZATION TRAINING

We derived our estimates from several sources. For those who support training directly, we modeled our requirements on the O/C structure at the NTC.[1] We

---

[1]NTC Operations Group (1994b).

also reviewed the active component support provided to ARNG brigades during Annual Training,[2] support provided to the reserves during the preparation for Desert Storm,[3] and active component training.[4]

We assumed that the conditions of mobilization would preclude any active component TOE units from assisting with the mobilization of reserve component units. We also assume that only some of the active and reserve personnel involved in the training of the RC during peacetime would be available. Others in this category would be fully engaged in mobilizing and training units assigned to the higher-priority reserve component CFP units.

The training model requires trainers external to the brigade being trained, training support personnel, and administrative and logistics personnel from the installation where the training takes place. Here we discuss the types of training support personnel required, our methodology for quantifying the requirements, and the support requirements of the training model. We next outline the potential sources for this support. Finally, we compare the requirements for trainers and training support with available resources.

Types of Personnel Required

We divided the functions that required training personnel into five categories: trainers, training management personnel, training support personnel, simulation support personnel, and installation and higher-echelon support personnel.

Trainers. These personnel observe and record the results of training and facilitate the After Action Reviews. They also control the exercises. Importantly, they mentor or conduct informal one-on-one training with the leader of the unit they observe. For maneuver training exercises, these trainers are called Observer/Controllers–Trainers (O/C-T). For gunnery exercises they are called Bradley or Tank Crew Examiners (BCE or TCE).

---

[2]2nd Armor Division (1993), 1st Cavalry Division (1993), 4th Infantry Division (Mech) (1993); 4th Infantry Division (Mech) (1992), and 24th Infantry Division (1992).

[3]5th Infantry Division (Mech) (1991b) and NTC Operations Group (1990a).

[4]2nd Brigade, 1st Infantry Division, *2nd Brigade Company Lanes Administrative Order*, Fort Riley, Kansas, December 1993; 1st Infantry Division, *Gauntlet 94-01 Exercise Directive*, Fort Riley, Kansas, December 1993; 1-8 Cavalry Squadron, *OPORD 94-09*, Fort Hood, Texas, February 1994; 2nd Brigade, 1st Cavalry Division, *OPORD 94-15, BLACKJACK Challenge*, Fort Hood, Texas, April 1994; 2nd Armored Division, *NTC Ramp Up Program*, Fort Hood, Texas, unpublished, January 1994; 4th Infantry Division (Mech), *Iron Point Lane Training Program*, Fort Carson, Colorado, July 1992; and 2nd Brigade, 24th Division, *OPORD 94-04 9 (Victory Focus I 94-03)*, Fort Stewart, Georgia, January 1995.

Trainers must be proficient, have field training experience in the job of the leader they are observing, should have undergone structured preparation, and have experience at performing O/C-T and BCE/TCE duties.[5]

Because of the high experience and preparation levels needed, we filled O/C-T and BCE/TCE trainer positions with AC leaders currently performing these types of training functions as a preferred option. However, experienced Guard personnel can perform BCE/TCE and even platoon-level O/C-T functions without significant training impact.

Training management personnel. This category includes all the personnel who plan and coordinate execution of the training events and who are not direct trainers. They develop training scenarios, write operations orders, develop training schedules, coordinate training area and range requirements, and manage lanes and gunnery ranges. Because these personnel must have high levels of experience early in the mobilization period, we sought to fill these requirements with AC personnel who perform similar functions in premobilization positions.

Training support personnel. This category of personnel includes all types of general support functions that normally accompany training. It includes such activities as guarding live-fire ranges, generating smoke, and marking fires. It also includes a small number of personnel who provide basic internal support for the trainers, such as their own maintenance and resupply. Training support personnel do not need to be as experienced as trainers and training managers. ARNG and AC personnel with the correct MOS could perform these functions.[6]

In addition, equipment is needed to support the training and the trainers. The equipment would include such items as vehicles, radios, tents, generators, and smoke generators. The amount of additional equipment needed would depend on what was already available within the training organizations and at the training sites. Although sizing this requirement falls beyond the scope of this study, training support personnel must be capable of maintaining this equipment.

---

[5]As an example, the O/C-Ts at the NTC are chosen based on successful completion of the job of the leader they "cover down on." They also attend a structured course at the NTC and normally spend a rotation understudying an experienced O/C-T before assuming their duties. For particularly demanding O/C-T positions (e.g., company commander or FSO), preparation normally includes serving in another O/C-T position initially to get a general understanding of the way training at the NTC is organized and conducted.

[6]Qualified USAR personnel could also perform these functions, but considering the requirement for providing corps-level support units and supporting their deployment and the availability of National Guard divisions to support this function for enhanced brigades, we have used ARNG personnel instead.

Simulations support personnel. These are the people who run and maintain the simulations supporting the preliminary command-and-control training. It does not include the O/C-Ts or training management for these training events; we include them under those categories. These personnel require experience in operating the simulations. There are many such personnel in the Division (Exercise) and at simulations centers at AC installations. We assume that they will be provided by the garrison staff and their contractors. If augmentation is required, USAR battle command staff training (BCST) personnel will provide it. Although these personnel would increase slightly the number required to support the training model, this support poses no problem, and we do not discuss it further or include it in our personnel requirements

Installation and higher-echelon support. This category includes all the organizations and personnel who provide outside administrative and logistics support to the ARNG brigade, OPFOR, and training personnel. Such support is provided by a combination of the TDA installation staff, contract support, and TOE support units assigned to the installations. Such support includes maintenance, medical, supply, finance and personnel activities, operation of ammunition and other supply points, and many other similar activities.

Methodology for Quantifying Requirements

To determine the number and type of training personnel needed, we examined doctrinal requirements outlined in Army training publications and AC and RC experience in conducting similar training. Requirements were determined for each training activity in the training model on each day based on averages of these data points. Training personnel requirements for each training event were calculated down to branch and grade level of detail. Although training support personnel were assumed available for subsequent events as soon as a training event was completed, we allowed a small amount of preparation time before each new training activity.

Given current plans, we next examined the potential number of training personnel who would be available to support ARNG brigade postmobilization training. In the analysis we took into consideration the competing priorities that would exist during such an intense mobilization period, such as training high-priority support units. Comparing available training personnel with the requirements allowed us to calculate the number of heavy brigades that could be trained at a single time.

To determine the amount of installation administrative and logistics support needed, we contacted the mobilization planning personnel at each proposed training site and asked them to estimate their installation's ability to support postmobilization training involving a full ARNG brigade with outside OPFOR.

We asked them to assume that they would maintain their current mobilization missions and that the brigades arrive at their installations soon after M-day. While we found no real limit on the number of enhanced brigades that could be trained at one time based on installation support, we did find that training the brigades would increase the number of units that would have to be mobilized, and that some installations would require more time to prepare for mobilization.

Requirements

Trainers, training managers, and training support personnel requirements. Using the methodology described above, we determined that it takes almost 1,000 people from these categories to support the training at a single company or brigade site. Table 4.1 lists the number of people required by category of training personnel.[7] Trainers and training managers have been gathered into one category because we try to fill these categories with experienced personnel, and the number of training managers is smaller than the number of trainers. Training support has been divided into two subcategories, lanes and range support and field support (e.g., maintenance, supply). Appendix C contains a more detailed explanation of the requirements.

Sources for Trainers

To determine the potential sources for trainers and training managers to support the training model, we considered the organizations that would have ex-

Table 4.1

Personnel Required to Support Training Site

Brigade Units	Number of People Required
Trainers and training management	637
Training support	
Lanes and ranges	231
Field support to trainers	99
Total	967

---

[7]We have not included the requirements for the air defense artillery battery and mechanized infantry detachment at this time. These organizations have recently been added to the enhanced brigades TOE. Our initial estimate is that these organizations will add a requirement for no more than eight trainers and possibly some control and support personnel to develop and coordinate OPFOR air activities. They could also add a requirement for OPFOR attack helicopters, which are currently not in that structure except at Fort Irwin.

perienced trainers who would be available during the postmobilization period. We first looked to active component sources and then to the reserve components. We generally provided training support personnel from RC units.

Active component support. During ODS, members of active divisions participated heavily in the preparation of the ARNG brigades called up. The 5th Division worked with the 256th Mechanized Infantry Brigade and the 4th Infantry Division with the 155th Armor Brigade. Today, however, the only AC TOE combat unit available to assist heavy enhanced brigade training would be the 11th Armored Cavalry Regiment (ACR) at Fort Irwin, California, and we use this organization as OPFOR rather than as trainers. We assume that all other active component combat units will have deployed.

The other reasonably available sources of AC trainers are those in TDA positions with premobilization missions of supporting the RC. The postmobilization mission of these personnel would be to assist the mobilization, readiness enhancement, and deployment of RC units, including enhanced brigades.

We evaluated each potential organizational source of trainers to determine their availability and to see if it appeared likely that they would have sufficient experience to perform in trainer and training manager roles. Potential sources we examined were the Combat Training Centers (CTCs) and the numerous AC organizations supporting the RC during peacetime.

Combat Training Centers. The Army's three Combat Training Centers have permanently assigned AC trainers who could support brigade training during postmobilization.[8] These are the Joint Readiness Training Center (JRTC) at Fort Polk, Louisiana, The National Training Center (NTC) at Fort Irwin, California, and the Combat Maneuver Training Center (CMTC) at Hohenfels, Germany.

For purposes of providing AC trainers for ARNG brigades, we concluded that only trainers from the NTC would be available. The JRTC trains light infantry brigades, and during a postmobilization period we assumed it would be training light enhanced brigades. The CMTC trains AC combat forces stationed in Europe. Because it is likely that the European divisions would deploy to a major regional conflict (MRC) before the enhanced brigades, we felt it unwise to assume that this training asset would be available. Given the fact that European divisions have one brigade each in the United States, have fewer field training

---

[8]The Battle Command Training Program (BCTP) at Fort Leavenworth is also considered a CTC but has a mission of training divisions and corps commanders and staff. We assumed that this organization would be involved in training for these echelons, as was the case during Operation Desert Storm. One of the BCTP teams is responsible for executing the brigade command and battle staff training program for ARNG brigades and battalions. However, this organization was formed from Title VII assets, and we discuss it later in this chapter.

opportunities than divisions in the United States, and will have to move through European ports, it is likely that at least some of these brigades will deploy after CONUS divisions. These later-deploying European brigades will be involved in intense preparation training that will most likely involve CMTC assets.

Approximately 750 trainers are assigned to the NTC Operations Group and stationed at Fort Irwin.[9] The NTC Operations Group conducts brigade-level maneuver and battalion-level collective live-fire exercises for CONUS heavy brigades. We assumed that NTC trainers would be available soon after mobilization. If an NTC rotation is in process, it would be completed and subsequent ones canceled. At that point, the NTC Operations Group would have the mission of supporting RC postmobilization training.

The trainers and training managers at the NTC have a strong comparative advantage in conducting brigade-level and battalion task force training. This advantage results from the fact that this cadre has a dedicated training mission, which they perform several weeks a month, twelve months a year. They are exposed to units across the Army, which gives them a unique opportunity to learn from the collective experience of units going through the training. Additionally, the NTC is the only location where brigade FTX and battalion CALFEX occur routinely.

Active component organizations supporting the reserve components. Approximately 7,700 AC positions are devoted to premobilization support of the RC. These organizations and the number in each are shown in Table 4.2.[10] We examined the availability and qualifications of each group to determine if they could serve as trainers in the postmobilization training model.

Units considered unavailable. We could not assume that some of these organizations would be available as postmobilization training assets, because they have other important postmobilization missions. **CONUSA, senior army advisors, USARC headquarters personnel, Third Army planners,** and **Information Systems Command** will be working in headquarters that support the overall planning, coordination, and execution of mobilization. The **Total Army Schools System (TASS)** personnel will be involved in postmobilization special and MOS schooling. **Full-time support personnel** are assigned to RC TOE and TDA organizations, most of which will either deploy or support the deployment

---

[9]NTC Operations Group (1994b).

[10]U.S. Army Forces Command, *Ground Force Readiness Enhancement OPPLAN, Draft,* Fort McPherson, Georgia, December 1994, and U.S. Army Forces Command, *Active Component to Reserve Component Support,* briefing, Fort McPherson, Georgia, March 1995.

Table 4.2

Possible Sources of Personnel for Postmobilization Training
(AC to RC)

Command/Unit	Number of Personnel	Needed for CFP or Light Infantry	Available for Heavy Brigades
FORSCOM			
CONUSA HQs	**169**		
Senior Army advisers	**108**		
Readiness groups (RGs)	**2,002**	2,002	
Field training groups—ARNG Divs	**80**		
ARMS Team	**11**	11	
Camp Dodge	**24**	24	
Regional training team/RTDs CFP	**464**	464	
Combat training centers	10		10
RTDs—14 CONUS enhanced brigades	655	271	384
Regional training brigades	1,820	485	1,335
ORE teams	146		146
USAR Div(E)s—5			
Division advisers	15		
Field exercise brigade O/C-Ts	**218**	218	
BCST simulation brigade O/C-T	170		170
TRADOC			
TASS RC schools	**415**		
BCBST Staff Simu (Fort Leavenworth)	62		62
MWST (Fort Knox)	19		19
USARC			
Full-time support	**947**		
Third Army planners	**171**		
ARNG			
Full-time support	**104**		
USARPAC			
ORE and RTD	**54**	54	
Information Systems Command	**10**		
Total	7,674	3,529	2,126

NOTE: Boldface indicates units we considered unavailable to support postmobilization training of ARNG brigades.

of other RC units. **USARPAC ORE** and **resident training detachments** (RTDs) would be supporting the deployment of RC units outside of the continental United States. **Field training groups** (FTG) are a small group of mainly senior leaders assigned to advise each National Guard Division. Given the possible support roles these divisions may provide, we did not use FTG as a source of trainers.

Readiness groups (RGs) are assigned to CONUSAs with the peacetime mission of assisting the training of RC units stationed in their area. Approximately 25 percent of the AC trainer support for postmobilization is in RGs. There are 29 RGs in the United States, with an authorized total strength of approximately 2,000. Each is commanded by a colonel and has an average peacetime strength of about 80 officers and NCOs.

The primary postmobilization mission of RGs is to form Mobilization Assistance Teams (MATs). MATs do not exist in peacetime; they form at designated mobilization stations when a significant number of RC forces mobilize.[11] They help mobilization station commanders and train and validate RC units for deployment.

We were not able to use current estimates of the availability of RG personnel for postmobilization training because no current plans define the number, grade, and MOS or the branches needed to meet MAT requirements at each mobilization station. However, we did examine the potential for using RG assets to support ARNG brigade training. We concluded that few have applicable MOS and that they would have demanding responsibilities to support the mobilization of earlier-deploying support units. We therefore do not use any to support the ARNG brigade's postmobilization training.

Validating and training RC CS and CSS units would be a large task for the approximately 2,000 RG and RTT trainers. Mobilization and deployment of the CS and CSS forces needed to sustain and support AC combat elements (even without ARNG brigades) will most likely require a massive effort. The proportion of RC forces in comparison to AC in the CFP is far larger than it was in Operation Desert Storm, there are fewer AC and RC forces to draw on, and it is possible that a future foe would not allow the deliberate build-up prior to hostilities that occurred during Desert Storm.[12]

In a major deployment, extensive RC support forces would be needed in theater in 90 days. To support eight divisions they include almost 800 units and 120,000 personnel. During ODS, units that deployed by air generally spent from 8 to 16 days in training and validation. Those that deployed by sea averaged a little over three weeks in training and validation. While RG trainers would not be required continuously, the approximately 2,000 personnel authorized in RG and RTT would be stressed to handle training and validation functions for this flow at up to 16 mobilization stations.[13]

---

[11]To assess the status of RGs and MAT team resources, we reviewed FORSCOM regulations and guidance dealing with mobilization and discussed mobilization plans, procedures, and status with mobilization planners at FORSCOM, all four CONUSAs, five installations, and with seven RGs.

[12]General Accounting Office (1992b).

[13]See Lippiatt, Polich, and Sortor (1992).

It would most likely take several weeks to complete this mission for each unit. Our study of RC support unit training during 1992 and 1993 showed that priority RC support units were evaluated as needing an average of 21 training days to meet deployment standards.[14] While the FEB and BCSTs of the Division (Exercise) could certainly assist in this effort, there are no plans for their use or even their mobilization.

This analysis suggests that mobilization requirements will consume RG assets. Support units vary greatly in the complexity of their wartime mission and current training status. Many also have significant shortages of qualified soldiers and will require some level of training as these replacement personnel arrive. Assuming that the typical unit (150 personnel) would require 5 to 8 trainers or validators, and that only half of the units (400) would require training support at one time, 2,000–3,200 trainers would be required. Given the difficulty of shifting personnel between mobilization stations, and likely mismatches between the types of trainers needed and types of units being trained at any point in time, it does not seem reasonable to assume that RG assets would be available.

Resident training detachments/reserve training teams (RTD/RTT) for CFP. These organizations have both pre- and postmobilization missions. In peacetime, they help CFP unit leaders plan and conduct training. Following mobilization, they assist units in preparing for deployment. Given the priority of the CFP, we could not assume that these personnel would be available to support enhanced brigade training.

Units considered available. ORE teams are also a potential source of trainers and training managers.[15] They have both AC and Active Guard/Reserve (AGR) personnel assigned. (We consider the AGR personnel as qualified and available, but do not show them in Table 4.2 because it addresses only AC assets. See Table 4.3.) Part of their premobilization mission is to conduct training exercises for high-priority RC support units, and this has some application to ARNG brigade CS, and CSS unit trainer and trainer management functions. ORE team availability for ARNG brigade postmobilization training is uncertain. While they have no defined postmobilization function, their premobilization mission

---

[14]See Sortor et al. (1994).

[15]See the following documents from U.S. Army Forces Command: *First U.S. Army Reserve Component Support Team TDA Personnel Allowances*, Fort McPherson, Georgia, March 1994; *Second U.S. Army Reserve Component Support Team TAA*, March 1994; *Fifth U.S. Army Reserve Component Support Team TDA Personnel Allowances*, Fort McPherson, Georgia, March 1994; and *Sixth U.S. Army Reserve Component Support Team TDA Personnel Allowances*, Fort McPherson, Georgia, March 1994.

is most aligned to CFP training and validation.[16] We used these personnel as supplementary sources of trainers for staff, CS, and CSS elements.

Each enhanced brigade is supported by a **resident training detachment (RTD)** of 48 personnel.[17] These personnel help brigade leaders to plan and conduct training, and they coordinate support from the affiliated AC division. Exact compositions vary between RTDs, but generally each heavy brigade RTD has one lieutenant colonel, two majors, and one sergeant major authorized for the brigade headquarters, a major, one or two captains, one maintenance warrant officer, and one sergeant first class authorized for each battalion, and a captain and a sergeant first class authorized for the armored cavalry troop.

RTD personnel are selected based on appropriate experience in AC. Their lieutenant colonels normally have not been battalion commanders but have served in field grade positions at battalion level. Their captains normally have been company commanders, and NCOs have served as platoon sergeants. This experience gives appropriate experience as staff trainers, platoon and company lanes trainers, and gunnery trainers.

Current policy for these organizations following mobilization is unclear. We assume they will stay with the brigade to which they are aligned during the entire postmobilization period to assist with their training or fill critical positions, and that RTD personnel may deploy with that brigade if it deploys.

Based on this level of experience and availability, we have allocated RTD personnel to staff, gunnery, and lanes training roles, but only for the brigades they associate with during premobilization.

There are six **regional training brigades** (RTB), and they are assigned to assist both combat and combat support units. Each brigade has a mix of tank, mechanized infantry, field artillery, air defense artillery, and combat engineer battalions.[18] The mix of battalions is based on the number and type of RC combat and CFP units in the RTB's area of responsibility. The combat units they support include both light and mechanized infantry as well as tank and armored cavalry units. Combat support units include CFP combat engineer, aviation, field artillery, and air defense artillery units. During peacetime, they provide

---

[16]We discussed the postmobilization missions with all CONUSA. A few ORE personnel are designated to perform CONUSA staffing functions, but the planned use for the majority is as an on-call augmentation to installation mobilization assistance teams (MATs).

[17]U.S. Army Forces Command, *Resident Training Detachment Personnel Allowance,* Fort McPherson, Georgia, July 1994.

[18]See the following documents from U.S. Army Forces Command: *First U.S. Army Regional Training Brigade Personnel,* Fort McPherson, Georgia, April 1995; *Second U.S. Army Regional Training Brigade Personnel,* Fort McPherson, Georgia, April 1995; and *Fifth U.S. Army Regional Training Brigade Personnel,* Fort McPherson, Georgia, April 1995.

gunnery and lanes training for enhanced brigades and CFP aviation and CS units.

RTBs normally have experienced personnel assigned to them. Each is commanded by a colonel who has been a battalion commander. The battalions are commanded by lieutenant colonels who have served as battalion executive or operations officers. Almost all majors assigned have served on battalion staffs, and all of the captains should have commanded companies. RTB personnel are generally less experienced than NTC Operations Group personnel with respect to battalion- and brigade-level training. No former brigade commanders are assigned to RTBs, and only the commander will normally have battalion command experience.

RTBs conduct platoon maneuver training exercises and gunnery through Table VIII. They also conduct artillery training exercises through battery level and may carry out some company-level maneuver training. However, they do not perform command-and-control training for battalion and brigade commanders and staffs. Personnel assigned will generally have limited brigade staff experience, and RTBs do not conduct battalion- and brigade-level command-and-control exercises during peacetime.

RTBs do not have all the types of training personnel required to carry out a complete maneuver brigade training program. There are no CSS personnel assigned, and there is a limited number of trainers of the type needed (Stinger) for air defense battery in the enhanced brigades.

Nor would all RTBs be available to assist with the postmobilization training of the ARNG brigades. Some of the infantry personnel would have to support the training of the enhanced light infantry brigades. Combat engineer, field artillery, and air defense artillery RTB battalions would have to give priority to CFP units. Given the limited planning that has gone on for CFP postmobilization training, it is not possible to determine the availability of CS trainers for enhanced brigades. However, given the relatively small requirement for the enhanced brigades in this area, we assume that field artillery, combat engineer, and air defense trainers of the appropriate MOS would be available.

RTB personnel should be especially well suited for gunnery, maneuver platoon, and field artillery battery lane roles but will have limited capabilities initially for performing battalion- and brigade-level command and staff training. With some preparation, these personnel should be able to support maneuver company lane training.

Two organizations, **Brigade Command and Battle Staff Training (BCBST)** and **Mounted Warfare Simulations Trainers (MWST)**, are fully available for postmobilization training of the enhanced brigades. The BCBST stationed at Fort

Leavenworth has the mission of providing JANUS- and BBS-supported command and staff training for National Guard combat brigades.[19] The MWST conducts platoon- and company-level SIMNET exercises and battalion-level command and staff training exercises.[20] This organization also conducts some platoon lanes training. Given their premobilization mission, the command and staff trainers of both these organizations should be effective brigade and battalion staff trainers, although they will probably lack command experience at these levels. Additionally, the personnel in the MWST could be effective company and platoon lanes trainers. We used them as a primary source of staff trainers and training managers.

Reserve component support. Current **National Guard** maneuver brigade training strategies emphasize gunnery, platoon collective maneuver skills, and CS and CSS proficiency at the company level. We assume that we will be able to use experienced leaders from ARNG combat divisions as a supplementary source for trainers for and some types of trainer positions (e.g., TCEs and BCEs), and we have used personnel from ARNG units as primary sources of training support personnel. These units would also have most of the additional vehicles, radios, and other equipment needed to support the training personnel.

USAR Divisions (Exercise) provide lanes training for RC CS and CSS units and command and staff training for all types of RC headquarters in peacetime. To perform this mission, they have two types of brigades: battle command staff training (BCST), which conduct JANUS and BBS exercises for RC combat, CS, and CSS command-and-control organizations,[21] and field exercise brigades (FEB), which conduct lanes training for CS and CSS units up to company size.

While these organizations are from the USAR, each has AC officers and NCOs assigned as trainers and staff officers. The numbers listed on Table 4.2 represent AC personnel.

The likely priority of FEB for CFP training during postmobilization means that the AC trainers from those units would not be available. However, the AC personnel from the BCST should be available and qualified to act as training managers and as command and staff trainers given their grades, branches, and overall premobilization experience. We used these personnel as supplementary

---

[19]BCBST, briefing, Leavenworth, Kansas.

[20]*Mounted Warfare SIM Trainers*, briefing, Fort Knox.

[21]U.S. Army Forces Command, *Active Component Staffing for Battle Command Staff Training (BCST) Brigades*, Fort McPherson, Georgia, September 1994.

sources of trainers and training managers. We also planned to use some of the RC personnel from the BCST as simulations support personnel.[22]

Meeting the Personnel Requirement

Our goal in meeting the training personnel requirements is to provide active component personnel as trainers and training managers. Training support personnel could come from either the active or the reserve components. The total requirement for three sites is 2,823.[23] With some exceptions, we found that there were sufficient active component personnel from those groups we have identified as being available to staff the trainer and training manager requirements for three brigade-level training sites, with the use of approximately 60 RC personnel as BCE/TCE. There are some shortages, which we will discuss later. Table 4.3 summarizes the number of personnel used from each source. A detailed description of the sources identified for each site appears in Appendix D.

Table 4.3

Training Personnel Sources to Staff Three Brigade Training Sites

Source	Number
Active component	
AC organizations supporting RC[a]	1,158
National Training Center	562
Reserve component	
ORE team AGR O/C-Ts	76
ARNG O/C-Ts	63
Range and gunnery support	625
Field support to trainers	198
Total filled	2,682
Total required	2,823
Requirements not filled	141

[a]These break out as follows: RTD, 138; ORE, 72; RTB, 876; BCBST/BCST, 50; MWST, 22.

---

[22]The postmobilization command-and-control training requirement for CFP headquarters units has not been determined, and postmobilization missions for the BCST are not specified. This may be a competing requirement. However, many of the AC trainers in the BCST are combat arms personnel, and they are organized to be able to conduct command-and-control exercises for combat brigades. Considering these factors, we have used AC personnel in BCST as a supplementary source of trainers. Only 23 of 170 AC BCST trainers available were used to support enhanced brigade training.

[23]We have assumed that one of the sites will be at Fort Irwin, which has a somewhat smaller field support requirement because of the infrastructure already present there (889 total versus 967 at the other two sites). Thus, the total requirement is for 2,823 trainers and training support personnel (967 + 967 + 889).

Active component. To meet the 2,823 requirement, we use 1,720 AC personnel, and 1,158 come from AC organizations supporting RC units. Table 4.4 shows the sources of the trainers and the number required by each category to staff three sites, again assuming that one site is Fort Irwin.

The ORE teams have 146 AC personnel available, and we use 72 active component personnel and 76 AGRs from the ORE teams. (Again, the table reflects only the active component personnel.) They serve primarily as functional trainers for battalion CS and CSS battalion staffs. Recall that the RTBs are not resourced to provide functional staff training because their primary peacetime focus is on gunnery and lanes training. There remain 74 active component soldiers and 143 AGRs from ORE teams who would be available for other CONUSA mobilization manpower requirements.

We use RTD personnel but assume they will be available as trainers only for their own brigades. The RTDs for the seven enhanced brigades have 384 people available. We use only those for the mobilized brigades, a total of 138. The remainder will stay with their units to continue support of premobilization training.

The RTBs along with the NTC Operations Group are the primary sources of combat arms trainers. Of the 1,335 people potentially available, we use 876. As we discussed earlier, although some of the RTB battalions are targeted to support CFP units (artillery, engineers, air defense artillery, and aviation) and light infantry brigade training, there is no current planning estimate of the number of RTB personnel required to support postmobilization training. We use two of the six headquarters available for two brigade sites. The NTC does not need this

Table 4.4

AC Personnel Available for Postmobilization Training of Heavy Brigades

Command/Unit	Number of Personnel	Required for Three Sites	Available But Not Used
FORSCOM			
ORE teams	146	72	74
RTDs: 7 enhanced brigades	384	138	246
RTBs	1,335	876	459
USAR Div(E)			
BCST simulation brigade O/C-T	170	23	147
TRADOC			
BCBST Staff Simu (Fort Leavenworth)	62	30	32
MWST (Fort Knox)	19	19	0
Total	2,116	1,158	958

type of support. We do not use any of the RTB air defense artillery battalions because they lack the type of trainers needed (Stinger); we do not use any of the aviation battalions because the ARNG brigades do not have aviation elements. We assume the remainder are available for training CFP units.

Of the 459 available but not used, about a third belong to RTB headquarters (38 personnel each). The remaining four headquarters would be available to support CFP and light infantry training. Fifty percent of the available RTB personnel were split evenly between engineer and field artillery training battalions within the RTB structure. The remaining unused personnel belong to the RTB tank battalions. This excess occurs because we were sizing the requirement for mechanized infantry brigades. Had we included armor brigades in the requirements, these personnel would have been required.

The active component O/C-Ts in the BCST simulation brigades and the BCBST at Fort Leavenworth are senior staff trainers, and have 81 people available. We use 50 (23 and 27 respectively), primarily to train brigade staffs. The NTC has a team for brigade staff training, but the RTBs do not. Consequently, additional brigade command and staff trainers are needed. We use personnel from only one of the BCST active component teams, and the remaining four would be available to support CFP or light brigade postmobilization command-and-control training.

We used 22 of the 33 personnel from the MWST simulation training activity at Fort Knox to provide trainers for the tank battalions. Again, if the requirement were sized for armor brigades, all of the personnel could be used.

In addition to the AC to RC units, we assume NTC personnel would be available. Based on our requirements, we use 562 personnel from the NTC operations group. We did not use the aviation team (Eagles) because the enhanced brigades do not have aviation. The light infantry training team (Tarantulas) was not fully utilized because the requirement was based on training a mechanized brigade, and many members of the team did not have an MOS that matched postmobilization requirements. Finally, our requirement was developed for sites other than Fort Irwin, and we only require sufficient personnel to support company-level CALFEXs, which is less than the requirement for battalion-level CALFEX. Fort Irwin can support a battalion-level live fire, and the NTC live-fire team (Dragons) would be fully used for this exercise.[24]

Reserve component. In addition to the 76 full-time AGRs from the ORE teams, we needed 63 trainers from the ARNG. These would primarily supplement ac-

---

[24]The NTC is now reorganizing its teams to support brigade-level maneuver and live-fire training exercises. In assessing the ability of the NTC Operations Group to support the training model, we used the current rather than the proposed TDA.

tive component trainers as BCEs and TCEs during the gunnery qualification phase of training. They could be made available from ARNG divisions with like-type equipment or from ARNG heavy brigades not yet mobilized.

The principal requirement for the ARNG is to provide lanes and gunnery range support personnel (625) as well as field support (198) to the trainers. A total of 823 would be required for these functions and, as before, could be provided by ARNG divisions or brigades.

Requirements not filled. Within those sources of training personnel we considered available during postmobilization, we could not fill 141 spaces for a three-site requirement. These shortages fall into two major categories: (1) active component captains or E8s with the proper MOSs to provide functional training for all company executive officers and first sergeants, and (2) specialty MOS trainers. The latter category includes trainers in the chemical (NBC and smoke), medical, intelligence, air defense artillery, and signal specialty areas as well as a shortage of artillery 13Fs needed as FIST trainers. Some of the signal, intelligence, and artillery shortages could be filled from the RGs if they were not fully committed to mobilizing the CFP. The remaining shortages seem small enough to be met by assigning individuals from units that do not deploy.

In summary, we could identify sources for all but 141 out of 2,823 positions required to support three brigade-level training sites: 1,720 from the active component and 962 from the reserve component, including 76 AGRs. Given the specialties of the personnel in the AC to RC structure, including the RGs and those at the NTC, there is a sufficient number of active component personnel to provide training at three—but only three—heavy brigade training sites.

Trainer and Training Manager Planning and Preparation

Planning activities start with development of training schedules geared to the capabilities of the installation on which the training will take place, the training level of the brigade, and the combat situation of the theater to which the brigade will deploy. Planning also includes selecting the areas where training will take place, developing scenarios and detailed plans for each training event, determining the exact trainer and training support requirements, and selecting the organizations or individuals who will perform trainer and training management functions.

Preparation activities involve the setup of the training areas and organizations, preparation of the personnel who will conduct and support the training events, and coordination of the schedule and necessary support with installation and other supporting organizations.

Meeting the training model's timelines and quality goals requires effective planning and preparation. The schedule in the training model assumes a fully functional training organization by M+18. Although time is built into the training model for retraining and maintenance, this was based on likely training shortfalls in the brigade, not the training organization.

Meeting the model's time and quality goals will require some of the planning preparation functions to be accomplished during premobilization if training is to start on M+18. The postmobilization training sites, training organizations, and the leaders to be assigned to key positions in the training organizations should be identified. A basic plan and standard operation procedures should be developed and practiced.

The amount of time necessary to plan and prepare for execution of the model will vary by the number of brigades that are initially trained but will be extensive, even for a single brigade. Prior experience in similar situations can provide some comparisons. The NTC Operations Group spent approximately six weeks preparing for the 48th Brigade during Operation Desert Storm and actually began preliminary planning months earlier. The 4th and 5th Infantry Divisions spent slightly less planning and preparation time, but because they began their training mission later than the NTC, they were able to leverage their efforts from the NTC's experience. During 1992 and 1993, AC divisions spent three to four weeks preparing for gunnery qualification and platoon lanes for summer ATs and began planning at least six months prior to AT. These examples were for established organizations with experience and established procedures for conducting these types of training missions.

Trainer and training management functions for a single brigade can be performed by the NTC's Operations Group and a couple of RTBs with minor reorganization.[25] Training more than one brigade at one time will require major reorganization, and this will increase the needed amount of planning and preparation over that needed for a single site. Training two brigades would basically require forming two expanded Operations Groups. The most reasonable solution would have half of the Operations Group at each site, augmented with the most experienced personnel available from other sources to reach required manning levels. Formation of a third site would require either greater dissipation of the Operations Group or assignment of that mission to an organization with limited experience at conducting this type of training.

---

[25]The NTC Operations Group, with some augmentation, conducts preliminary command and control, company team, battalion task force, and brigade-level maneuver and CALFEX training; an organization formed from RTBs conducts gunnery, platoon, and support company–level training.

INSTALLATION SUPPORT

The ability of installations to support mobilization is the second area to examine. Each training site we examined deploys AC units and mobilizes and deploys a number of RC units. The minimum requirement is to get these units to a readiness level acceptable to the in-theater CINC, but a larger goal is to reach the highest possible readiness levels prior to deployment.[26]

Types of Support Required

This goal requires personnel, medical, maintenance, supply, and training support. Members of RC units must be administratively transferred into the active army and administratively and medically prepared for deployment overseas. If the mobilizing unit is short the required number of personnel, replacements must be identified and transferred into the unit. Equipment must be brought to mission-ready standards, and if needed equipment or supplies are missing, these must be ordered and transferred to the unit.

Installations also support any needed premobilization training. After the unit reaches deployment standards, the installation commander must validate that the unit has reached these standards and is prepared for deployment. Finally, the installation must arrange transportation to the port and theater.

Active component installations support similar activities on a day-to-day basis for the units stationed there. This support is provided by a combination of a TDA garrison military and Department of the Army civilian staff, contract services, and TOE units assigned to the installation. The exact combination varies at each installation.

To assist with the greater workload associated with mobilization, each garrison designated as a mobilization station develops a mobilization TDA to handle the larger postmobilization requirement. As a part of this mobilization TDA, installations get a USAR garrison unit and a USAR TDA augmentation to the installation's medical activity. The mobilization TDA also authorizes added DA civilian positions and officer and enlisted positions to be filled by personnel from the IRR or retirees recalled to active duty. Finally, most installations have in place postmobilization contract agreements to provide additional needed services from the civilian sector.

---

[26]Readiness levels are reported in accordance with AR 220-1. The criterion for a deploying combat unit during Desert Storm was C-1, "able to undertake the full wartime missions for which it was organized." For CS and CSS units, the criterion was C-3, "able to undertake many, but not all of the missions for which it was organized." The theater commander could accept a unit with a lower rating. Naturally, efforts were made to get units to higher levels of readiness if resources were available.

Methodology for Determining Requirements

Installation support requirements for enhanced ARNG brigade postmobilization training have not been identified, so we developed our own estimates. To estimate the level of installation staff support needed, we queried the mobilization planners at the six installations we had chosen as potential brigade training sites. In all cases, they thought they would require at least major portions of their mobilization TDA, to include their USAR garrison unit and TDA medical augmentation. Additionally, they would need RC TOE support units to perform maintenance, transport, supply, and other necessary support for ARNG brigade field and gunnery training not covered by their current planned augmentation.

To approximate the requirement for additional installation support, we assessed the personnel in TOE units that support current AC divisional installations with two or three heavy maneuver brigades.[27] The assumption is that these units are the minimum necessary in addition to their current TDA garrison and medical activity staffs. Our estimate is that, assuming its current premobilization TDA garrison and medical activity, the typical AC divisional installation requires approximately 1,700 additional personnel. These may be provided from either RC TOE units, RC TDA organizations, DA civilian hires, or contract support.[28]

Requirements

Table 4.5 reflects our estimate of the requirements in addition to permanent garrison TDA and medical detachment for a typical brigade site (one ARNG brigade and one brigade of OPFOR).

Installation support is critical to maintaining the optimistic timelines in the model for the initial set of brigades. Unless adequate installation support is available soon after M-day to maintain the relatively high training tempo, the time to execute the training model will increase, perhaps significantly.

With respect to maintaining the timelines in the training model, AC installations that support AC divisional- or brigade-level training during premobilization have an advantage for the initial set of brigades. The amount of their garrison augmentation requirement is reduced in that they have an established sup-

---

[27]We used Forts Carson, Riley, Stewart, and Irwin as the basis for our estimate. Appendix E contains a more detailed estimate. See also U.S. Army Forces Command, *Mobilization Station Study*, Fort McPherson, Georgia, February 1994.

[28]This estimate may be low. The postmobilization OPTEMPO of the enhanced brigades is likely to be higher than current AC peacetime rates, there will likely be additional medical, personnel, and finance activities given the nature of mobilization, and there will be less availability of borrowed manpower from TOE units.

Table 4.5

Installation Support Requirements

Unit Type	Number of Personnel
Adjutant General	
AG company	129
Finance battalion	84
Public affairs team	5
Medical	
Medical detachment	151
Sanitation detachment	10
Dental	23
Air ambulance	130
Military Police	
CID detachment	11
Combat support company	177
Supply	
Combat support company	134
Transportation	
Light/medium truck company	117
Movement control detachment	7
Ordnance	
EOD team	17
Maintenance	
Nondivision maintenance company	235
GS maintenance company	218
ATE repair detachment	7
Signal	
MSE battalion (2 nodes)	280
Total	1,735

port system. Similarly, Fort Irwin's premobilization mission and installation could probably allow it to transition quickly to support postmobilization. Fort Bliss would have somewhat greater problems in this area. However, establishing an effective installation support system early in the mobilization period for Gowen Field and Yakima would be very difficult, given the lack of current AC installation support.

Sources for Installation Support

A wide range of sources can supply the type of support needed at the installation level: nondivisional units that do not deploy, ARNG division support commands, mobilization TDAs, and contractors.

Nondivisional support units that do not deploy. Although we anticipate that all of the active component divisions would deploy, a number of support units are assigned to organizations above the division structure, and some of these may not deploy. Some RC support units would be available.

ARNG division support commands. The ARNG structure contains several divisions, each of which has a support command. These divisions are not required to meet current strategic requirements and would deploy late, if at all. Therefore, their support commands could provide much of the support required at the installation level.

Mobilization TDA expansion. For those types of support that depend more on individual than collective skills (e.g., medical), a mobilization TDA could prove the most effective way of providing needed installation support.

Contractors. Commercial contractors can provide many of the services needed (e.g., transportation). Contracts could be negotiated prior to mobilization and then activated on or close to M-day. The support could be in place by the time the brigade arrived at the training site.

Simulation Support

Simulation support can come from two sources. All AC installations have some support available. Furthermore, the BCBST cadre at Fort Leavenworth and the tactical simulation trainers at Fort Knox have contract simulations support that would be available during the postmobilization period and could be drawn on for support. Additionally, the BCST of each USAR Division (Exercise) has simulations personnel who can supplement these sources. The combination of the installation, BCST, and TRADOC personnel can more than satisfy the simulation support requirement for three brigade sites.

Mobilization Requirements

In the preceding chapters, we have discussed the need for ARNG personnel to fill some trainer positions and most training and field support requirements. We have also identified the requirement to replace some corps support group units at each installation to provide full garrison support, as they will most likely be deployed. These latter requirements could be filled by mobilizing reserve component units, expanding the installation contractor base, or hiring additional DoD civilians.

In the worst case, where the all support functions would have to be assumed by reserve component units, manning three brigade sites with training and garrison support would require the mobilization of about 6,100 personnel. This represents about 2,000 per site, although this number would vary from site to site depending on its current garrison structure and in-place contractor support.

TRAINING SITE COMMAND AND CONTROL

The quality of the training depends on far more than simply the sum of the qualifications of the individual trainers and training managers. The quality of training and the time required to execute it are the results of an effective training and support organization. This organizational effectiveness requires a cohesive organization with established and practiced procedures. Establishing such an organization requires time, effort, and an effective command structure.

As we have discussed in the preceding sections, staffing a brigade-level training site will draw personnel from a diverse set of sources that have little peacetime command relationship. Training sites would have four different commanders, each with different responsibilities and interests. The installation would be commanded by a senior colonel garrison commander, who has a garrison staff augmented by a mobilization TDA. Additionally, each installation has a mobilization assistance team (MAT) commanded by a colonel and a number of reserve component support units. Also present would be one or more senior colonel RTB commanders with their organizations; an OPFOR, again commanded by a senior colonel with his OPFOR brigade; and, finally, a brigadier general, there to train his enhanced separate brigade.

We believe that a general officer supported by a small staff is required to command each site to accomplish such activities as overseeing and directing training execution; coordinating training activities between the unit being trained, the installation, and the training organization; and interfacing with CONUSA, FORSCOM, and the TAGs of the enhanced and OPFOR brigades' home states. Such a command structure is in place at Fort Irwin, but if three brigade sites are required, two more such commands will be needed. The recent drawdown and the Army's restructuring of its divisions will leave two major generals with staffs at Fort Riley and Fort Carson whose peacetime responsibility is to command the fort and oversee the training of an ACR, three brigades belonging to divisions not located at those posts, and various other active component units located at those posts. These generals and their staffs could be made available to oversee the training at two sites since they would not deploy with the active component units under their peacetime supervision.

Previous discussions have emphasized the importance of thorough peacetime planning and preparation to assure a smooth transition to postmobilization training. We believe it is important that the general officers and their staffs play key roles in that planning and preparation. This would require these generals to be, in a sense, dual-hatted in peacetime. They are responsible for their peacetime mission through the normal FORSCOM chain of command. For their postmobilization mission, they would be responsible through the CONUSAs. The required peacetime planning and preparation will take a considerable effort, and existing staffs may have to be augmented with additional resources.

SUMMARY OF TRAINING AND SUPPORT PERSONNEL

Our analysis shows that five categories of training personnel are required, and that two of these categories—trainers and training managers—should be filled with the most experienced people available. Although shortages occur in certain categories of personnel, generally enough training personnel are available from our identified sources to train as many as three brigades at a single time. But trainers, managers, and training support personnel become the binding constraint in any effort to expand the training sites beyond three. Splitting the NTC's assets between two sites poses no particular challenge, but expanding to the third site raises the risk to both timelines and quality.

Furthermore, support beyond that devoted directly to training is required at the installation. Slightly over 1,700 people per installation are needed for administrative and support functions. Adequate sources exist to provide this support. Table 4.6 summarizes the training and installation support requirements for one and three sites.

Some trainer shortages result from not having personnel of the right grade, branch, or MOS available in the organizations considered available. The num-

Table 4.6

Training Personnel and Installation Requirements

Type of Personnel	Typical Site	Three Sites (one NTC)
Trainers and training management	637	1,911
Training support[a]		
Lanes and ranges	231	714
Field support to trainers	99	198
Installation augmentation	1,735	5,205
Total	2,702	8,028

[a]NTC requires a different mix of training support personnel; training support is provided by the garrison, but additional personnel are required to man the Tactical Analysis Facility that supports the instrumentation at Fort Irwin.

ber of these shortages is small enough that it seems feasible to replace them using existing individual replacement mechanisms.

Establishing three brigade sites initially without extensive premobilization planning and preparation poses both qualitative and timeline risks. No peacetime organization is sufficiently sized to meet the trainer and training manager requirement. The NTC Operations Group is closest, but even it requires augmentation. For two sites, and far more for three sites, the trainers will have to come from many different organizations that have no current peacetime relationship. New training organizations will have to be created, and complete planning and preparation for a very complex mission must be done in a very short time.

Finally, the issue of installation command and control must be addressed. Fort Irwin has an organic command-and-control structure with a general officer and a supporting staff. Other installations need something similar, and the commanders and staff at Forts Riley and Carson could fill this need.

Chapter Five

OPFOR

Our training model requires a qualified, dedicated OPFOR. The OPFOR must be of sufficient quality to meet training objectives. Current Army training doctrine directs MILES force-on-force training as a fundamental requirement for maneuver unit training exercises.[1] Additionally, a dedicated OPFOR is required to meet the timetable. In our model, the brigade being trained has no responsibility for providing OPFOR. The OPFOR is basically an orchestrated training aid that provides the required support when and where needed with no attempt to train the OPFOR unit. Availability of a dedicated OPFOR allows the brigade to move through the training model at an accelerated pace because the unit being trained does not have to detail any of its units to the OPFOR missions at the expense of its own training.

This chapter describes our methodology for determining the OPFOR requirements, identifies available resources needed to meet these requirements, and discusses the issues of quality and command and control. The chapter also addresses the need for quality in the OPFOR and some sustainability considerations.

METHODOLOGY FOR DETERMINING OPFOR REQUIREMENTS

OPFOR requirements have two dimensions. The first is a relatively straightforward numerical determination of the number of combat vehicles and dismounted infantry needed to support the training events, and from this the amount of units needed to provide these numbers of personnel and equipment.[2] The second pertains to quality and command and control. OPFOR support involves more than simply parceling out combat vehicles and infantry

---

[1] For example, see ARTEP-MTPs 71-1 and 71-2. Full citations for these documents can be found in the Bibliography under the heading "Army Training Documents," Department of the Army.

[2] These requirements were derived from the Krasnovian Doctrine and organizations as practiced by the OPFOR at the NTC.

to training sites. A command-and-control system is required for both tactical and administrative reasons. The command system is necessary to ensure that the right number of vehicles and soldiers are ready to perform their missions at the right place and time. It also ensures that the OPFOR organization conducts its operations to high tactical standards. Quality of the OPFOR means that it is tactically proficient and accurately reflects threat doctrine.

To determine the ability to support OPFOR requirements of the training model, we used a three-step process:

- Size the requirement by determining how many units are needed to provide tanks, Infantry Fighting Vehicles (IFVs), dismounted infantry, and other combat vehicles to support the training events in the training model.

- Identify the resources available to support this requirement and the number of brigade sites that could be supported at one time.

- Examine the feasibility and supportability of the OPFOR requirement from a command-and-control and OPFOR quality perspective.

SIZING THE OPFOR REQUIREMENT

To determine the number of combat vehicles (tank, BMP, reconnaissance, and combat support) and dismounted infantry needed, we drew on several data sources. We reviewed the OPFOR requirements listed in ARTEP-MTPs for the collective tasks selected for training in the training model.[3] We also examined OPFOR mixes used for similar AC and RC training events, including the NTC's OPFOR for battalion and brigade operations, AC experiences in preparation for the NTC, and AC support provided to National Guard brigades for Desert Storm and for 1992 and 1993 Annual Training. The OPFOR to BLUEFOR ratio we developed is generally less than that suggested in ARTEP-MTP, but it aligns with the data from AC and RC training events. Based on discussions with and review by numerous Army trainers, we believe it is sufficient to meet training objectives. We kept a relatively large amount of dismounted infantry and combat engineer strength in the OPFOR organization to give it the flexibility to portray different threat forces.

We made no attempt to replicate all OPFOR combat support, combat service support, or command-and-control elements, but established requirements only for those OPFOR elements necessary to reach the objectives for each training event. For example, we allocated no OPFOR field artillery or mortar vehicles to

---

[3]Included were ARTEP-MTPs 71-1 (Company Team), 71-2 (Battalion Task Force), 7-8 (Infantry Platoon), 17-237-10 (Tank Platoon), and 17-57-10 (Scout Platoon). These documents are fully cited in the Bibliography.

platoon and company lanes. While these exercises involve indirect fires, these weapons themselves would not normally be seen by platoons or company teams. On the other hand, engineer and command-and-control vehicles that would be seen were allocated to these same lanes.

The amount of OPFOR needed to support the company lanes and the battalion and brigade operations phases of the training model appears in Table 5.1. This table shows only the company team lanes and battalion and brigade operations because platoon lane training in all cases requires less support than these higher-echelon training events.[4]

IDENTIFYING THE NUMBER OF UNITS NEEDED FOR OPFOR

We next identified sources for the tank, BMP, other combat vehicles, and dismounted infantry requirements for the training model. Potential sources consist of the 11th ACR (OPFOR at the NTC) and ARNG combat units from other than enhanced brigades.

The type and number of organizations required to provide the combat vehicle and dismounted infantry requirements can be determined by comparing the vehicle or infantry requirement with the number in ARNG battalions and brigades and in the 11th ACR. Table 5.2 shows these numbers. Comparing the requirement with the assets available in the 11th ACR and an ARNG mechanized infantry brigade shows that all requirements can be met by augmenting the 11th ACR with one ARNG mechanized infantry battalion reinforced by additional dismounted infantry. The 11th ACR would thus require ARNG augmentation of about 1,000 people to support training a brigade with three maneuver battalions. Each of the other brigade-level training sites would require

Table 5.1

OPFOR Vehicle and Infantry Requirements

	Tanks	IFVs	Other Combat Vehicles	Dismounted Infantry
Company team lanes	36	97	20	429
Battalion task force and brigade operations	53	172	89	430

NOTE: The highest requirement for all categories is battalion task force and brigade operations. Other vehicles include those for reconnaissance, engineer, air defense artillery, and field artillery.

---

[4]A more detailed discussion of OPFOR requirements is presented in Appendix F.

Table 5.2

Available OPFOR Combat Elements from the 11th ACR and a National Guard Mechanized Infantry Brigade

Requirement	Tanks	IFVs	Other Combat Vehicles	Dismounted Infantry
Company team lanes	36	97	20	429
Battalion task force and brigade operations	53	172	89	430
Source				
11th ACR	40	129	70	0
ARNG mech infantry brigade	58[a]	120[a]	100[b]	432[c]
ARNG mech infantry battalion	0	64	20	216

[a]These figures are for a modernized mechanized infantry brigade with one tank battalion (58 tanks), two BFV mechanized infantry battalions (64 BFVs, 216 dismounted infantry each) and one cavalry troop (9 tanks, 8 BFVs).

[b]Includes mortar and engineer M113, field artillery battalion M109, and a portion of the battalion's HMMWVs.

[c]Based on 48 nine-man squads from the two mechanized infantry battalions.

an OPFOR organized as a separate brigade with its division slice and an additional mechanized infantry battalion, about 5,000 people (see Appendix F). Three sites—one being Fort Irwin—would thus require the mobilization of 11,000 ARNG personnel (1,000 + 5,000 + 5,000). A company training site would require only 3,600.

COMMAND-AND-CONTROL CONSIDERATIONS

Effective OPFOR operations require command-and-control proficiency. Leaders must be tactically knowledgeable and proficient at controlling their units and employing combat support assets. Units must have collective skills to react responsively to changing tactical situations and have mastered basic field operating procedures.

Command-and-control proficiency is difficult to achieve. Our studies show that AC units seldom reach desired levels of performance at the NTC, even after focused training programs.[5] Army leaders stress the difficulty of achieving proficiency in this difficult function. Career-long development of individual leaders' technical, tactical, and leadership skills and extensive collective training of units are required for units to develop effective command and control.

---

[5]Grossman (1994).

The OPFOR has to have a high level of command-and-control proficiency earlier in the postmobilization period than do the enhanced brigades. OPFOR battalion-level proficiency is required for company team lanes, and brigade-level proficiency is required to support battalion operations. It is not reasonable to assume that ARNG leaders from combat units with lower priority than enhanced brigades can reach full proficiency at battalion and brigade level soon after mobilization, given the limited training time and resources available during peacetime.[6]

The 11th ACR is proficient at command and control. Moreover, its leaders are expert in the OPFOR tactics. It has reached high levels of proficiency because it conducts OPFOR operations against BLUEFOR units during NTC rotations approximately 170 days a year—far more field training time at a higher echelon level than any other army unit.

The 11th ACR has a chain of command capable of commanding and controlling two brigade-sized motorized rifle regiments (MRRs) and eight battalion-sized motorized rifle battalions (MRBs). This is because each of the battalions in the 11th ACR alternates the role of commanding and controlling the MRR and its four MRBs for an NTC rotation. Thus the 11th ACR with National Guard augmentation could, with some preparation time, perform OPFOR roles at one or two brigade training sites with little if any degradation in command-and-control skills by using each of the battalions of the 11th ACR as the tactical chain of command at both sites. However, fielding more than two sites would be far more difficult, spreading the 11th ACR leadership cadre very thinly across the sites. The current chains of command would have to be reorganized, and many of the senior leadership positions would have to be filled with National Guard leaders who would have far less time to achieve the level of proficiency of the 11th ACR's full-time leaders.

QUALITY CONSIDERATIONS

A more subjective consideration than the number of sites that can be supported is the assessment of the ability to provide quality OPFOR and the relationship of a quality OPFOR to the training readiness level of the brigades being trained for deployment. We interviewed a number of Army leaders on the topic of OPFOR quality, including the commander of the NTC, two former commanders of the OPFOR, and two former commanders of the Operations Group. The majority

---

[6]For example, under the current funding plan, ARNG division units will have less funding for additional training days, less full-time manning, less priority for schooling, less access to BCBST exercises, less AC support, and smaller authorized strength levels than enhanced brigades. Such resource constraints will generally limit training attainment levels in comparison to the enhanced brigades.

consider an effective OPFOR to be a critical component of successful combat unit training. They credited the CTCs as being instrumental in developing the training readiness levels leading to the success of U.S. Army during the operations in Panama and the Persian Gulf War. This majority felt that the highly competent OPFORs at the NTC, CMTC, and JRTC were a major basis for their success.

However, during the course of our research, some of those we interviewed questioned the need to have a unit dedicated to performing the OPFOR function. An assumption we have made is that the brigades must reach "fully trained" status prior to deployment. This was the criterion established during Desert Storm. The current Army standard for preparing units and leaders for combat includes training at the NTC with its effective, dedicated OPFOR. Since the inception of the NTC, Army leadership has sent a regimental headquarters, a tank battalion, a mechanized infantry battalion, and a support battalion to the NTC to ensure effective training there through the use of a tactically correct OPFOR. We see that reaching "fully trained" training status would normally include the experience of training against an equivalently proficient OPFOR.

The quality of OPFOR has three important aspects. First, the OPFOR must be tactically proficient. It must be so proficient that all the tactical weaknesses of the BLUEFOR will be exposed. The BLUEFOR should be tactically successful because it is proficient, not because the OPFOR made mistakes. The second aspect is that OPFOR must be at the right time and place so that the training schedule is maintained. Moreover, it must be flexible enough to react to changes in schedules or to reexecute missions if additional training is needed. Third, the OPFOR should as far as possible replicate the tactics of the enemy so that the BLUEFOR develops effective tactics and techniques to defeat him. This is more than an abstract concept for the circumstances under which the enhanced brigades would perform postmobilization training. Under these circumstances the enemy would be known and the OPFOR organizations and doctrine appropriately modified, as was the case during Desert Storm.[7]

Expanding the OPFOR beyond the 11th ACR and maintaining quality standards will require pre- and postmobilization training for the ARNG units designated as OPFOR. The extent of the training requirement will be a factor of the amount of expansion and the training level of the OPFOR at the time of mobilization.

We discussed expanding the OPFOR after mobilization with numerous OPFOR leaders. They were reasonably confident about the potential for expansion to two brigade sites. National Guard units had frequently augmented the OPFOR, and those experiences had been generally positive. They were confident about

_____
[7]It took the NTC OPFOR about 60 days to develop its new Iraqi doctrine (Sumerain).

the ability of the 11th ACR and a Guard brigade to expand, given the time available for training during the lanes and gunnery phases.

However, expansion requires a premobilization program to prepare the 11th ACR and the National Guard augmentation elements for expansion. The following are elements regarded as necessary for effective expansion.

- The National Guard brigades' entire premobilization federal mission must be preparing for the OPFOR postmobilization mission.

- The OPFOR mission must be practiced during IDT and Annual Training. The National Guard units could rotate between missions of:

  — Providing OPFOR support for company- and platoon-level lanes for other Guard combat units during AT.

  — Augmentation at company level and below at the NTC during active component unit training. Individuals could perform individual augmentation at battalion level and above—ideally annually.

- The 11th ACR should be the associated AC unit for the Guard units with an OPFOR mission. It would support the AT and IDT training of the OPFOR as well as approve their training plans.

- Each National Guard brigade OPFOR would need a resident training detachment (RTD) of approximately the same size as the ones assigned to enhanced brigades. The 11th ACR would also require a small planning and coordination cell.

- The 11th ACR must conduct premobilization reconnaissance and coordination at training sites.

- OPFOR postmobilization plans must be developed in detail.

- The 11th ACR OPFOR cadre and National Guard OPFOR units must be available for training and preparation at the training site not later than training day 12.

These leaders were less optimistic about the ability to expand the OPFOR to support three sites and maintain current quality levels. They believed that the 11th ACR would have difficulty maintaining an effective premobilization training relationship with more than the ARNG origination needed for two brigade training sites. They expressed concern about the ability of an RC organization to command and control an OPFOR MRR (plus tanks) so soon after mobilization and for their own organization be an effective cadre at more than two sites.

BLUEFOR as OPFOR

We have been asked to consider options that would allow use of one ARNG enhanced brigade to conduct force-on-force training against another brigade while both are preparing for deployment. Such an option would have no dedicated OPFOR. Instead, one element opposes another, and each functions as the OPFOR for the other. For example, one brigade could be training the mission of deliberate attack against a battalion of the other brigade that was training the mission of defend in sector. Such a use of two organizations opposing one another during force-on-force training is often called BLUEFOR-on-BLUEFOR, and most AC organizations use this method for home station training.

We examined how a BLUEFOR-on-BLUEFOR concept would work in our training model and found that it is feasible and would reduce the amount of OPFOR support required. However, use of such a concept would almost certainly reduce the training quality, increase the time required to execute the model, or, most likely, both.

The current AC standard for maneuver unit training includes the experience of training against a well-trained OPFOR. Home station training programs are only a part of a total training strategy for AC heavy brigades because all also conduct tactical training at either the CMTC in Germany or the NTC in CONUS. While it can be argued that in the event of conflict, many AC brigades will deploy without a recent CTC rotation, under current rotation and personnel assignment policies, all AC units will still have a high portion of leaders who have trained against a high-quality OPFOR, if not with their current unit then with a previous one. ARNG brigades would have a very small percentage of leaders with such experience.

The training quality would probably decline for two reasons. First, both forces would use U.S. tactics and organizations. Thus there would be far less chance to gain an appreciation for the enemy's tactical strengths and weaknesses and to gain experience at recognizing and reacting to the tactical situation. Second, BLUEFOR-on-BLUEFOR training would be conducted as free play, and thus one side or the other could make mistakes. When a structured OPFOR is used, its plan and tactical dispositions are orchestrated to reduce OPFOR mistakes to a minimum. Rehearsals can increase the likelihood of reasonably high performance by a dedicated OPFOR.

The time to execute the model would probably increase. The model provides only limited time for retraining. When BLUEFOR-on-BLUEFOR methods are used, retraining requirements can be generated by two sides rather than one. The chances of the timelines being missed because of training mistakes (e.g., missing the time to cross the line of departure) would also increase.

Thus we believe that use of a BLUEFOR-on-BLUEFOR concept, although feasible, would not achieve the model's objectives for quality and time.

SUSTAINABILITY CONSIDERATIONS

The OPFOR organizations require sustainment support in the areas of maintenance, supply, personnel services, and MILES. Each of the National Guard Brigades has a forward support battalion. However, unlike the support battalions in the enhanced brigades, these battalions are not designed to support the brigade completely and require additional support from the division sustainment base.

An additional issue is the lower priority of these organizations. CFP and enhanced brigades have priority for personnel and equipment fill, schools, repair parts, funds, and other resources. It is likely that an ARNG brigade used in an OPFOR role would need considerable support from its parent division and assets from other brigades in the division to reach readiness levels to perform the OPFOR mission.

SUMMARY OF OPFOR

A proficient and dedicated OPFOR is key to successful implementation of the training model. Alternatives such as BLUEFOR-on-BLUEFOR pose unacceptable risks to the quality and timeline objectives of the model. Effective OPFOR support for two sites seems reasonably achievable; supporting three sites has a much higher level of risk from a perspective of C2 and quality.

Sufficient sources exist to fill the OPFOR requirement for two or three sites. A fully manned, reinforced ARNG mechanized infantry brigade has enough combat vehicles and infantry to provide an OPFOR for a brigade-level training site. The NTC would require augmentation of about 75 vehicles and 1,000 people. Additional sites would require a full brigade. However, given the lower priority of these brigades, all would need additional reinforcement and sustainment support, with the most available source being its parent division.

The aspects of OPFOR C2 and quality appear to be the greater limitations on the achievable number of brigade training sites. The consensus of the OPFOR-experienced leaders with whom we discussed this issue was that serious OPFOR quality issues would arise if more than two brigade sites were manned at one time. If this view is correct, the timetable or quality objectives of the model would be at risk of not being met for more than two brigade sites.

Even expansion to two brigade sites requires the ARNG units to assume the OPFOR mission and a primary activity and the implementation of a training relationship between the 11th ACR and the ARNG units that will perform OPFOR functions during postmobilization. Creation of a small AC coordination and support organization to assist premobilization OPFOR training is also a probable requirement. Additionally, the Guard units designated as OPFOR must focus on preparing to execute this mission during premobilization if they are to be proficient soon after mobilization.

Chapter Six

OPTIONS FOR EXECUTING THE TRAINING MODEL

Previous chapters have described the training model, the resources it requires, and the training locations and personnel sources to meet those resource requirements. Although five sites could support the training, we have identified sufficient sources of active component trainers and training managers to support only three brigade training sites. We have also identified the need for ARNG units to augment an 11th ACR OPFOR cadre at each site and concluded that supporting two sites would allow the 11th ACR to maintain critical OPFOR command and control and to provide a critical mass of highly qualified AC OPFOR at each site. Providing OPFOR to more sites raises command-and-control problems and diminishes the size of the 11th ACR cadre at each site in ways that may reduce the training quality. However, we conclude that it is feasible to establish up to three postmobilization brigade-level training sites, accepting some degree of increased risk to the training quality and timelines at the third site.

But the fact that the Army can establish three sites does not necessarily mean that it should. The determination of how many sites to establish and what training events to execute at each site will result from policymakers' analysis of tradeoffs along three dimensions: risk, resources required, and force-generation rates. Risk has two aspects. The first is training quality, and the second is achieving the training model's timelines, which ultimately affect the force-generation rate, that is, how quickly brigades can be mobilized, trained, and deployed.

To provide a framework for this analysis, we have developed options we believe can be executed with available resources and, given the assumptions noted in previous chapters, meet the quality and timeline objectives, albeit with some differential in risk.

This chapter discusses three alternatives for executing the model. One calls for all units to pass through a single brigade-level training site at Fort Irwin; the other two each add one more brigade-level training site. It then assesses these options along the three dimensions mentioned above. We note that, in general, as the number of sites increases, so do the force-generation rates, the resources required, and the risks.

OPTIONS

Figure 6.1 shows the three different options.

Option 1: One Brigade Site

Under this alternative, three training sites are established. A brigade training site is at Fort Irwin, and two company-level training sites are at other installations. All NTC resources, trainers and OPFOR, remain at Fort Irwin, and each ARNG brigade goes there for battalion live-fire (CALFEX) and battalion task force and brigade-level maneuver training. To improve the rate at which ARNG brigades can be trained in this alternative, two company-level sites are used to provide gunnery qualification, preliminary command and staff training, and platoon and company lane training. These activities represent the first 67 days of the training model described in Chapter Two (refer to Figure 2.1 and Appendix A).

To ensure a smooth flow, starts are staggered both at and between company training sites. Figure 6.2 illustrates the concept. A new brigade arrives at each

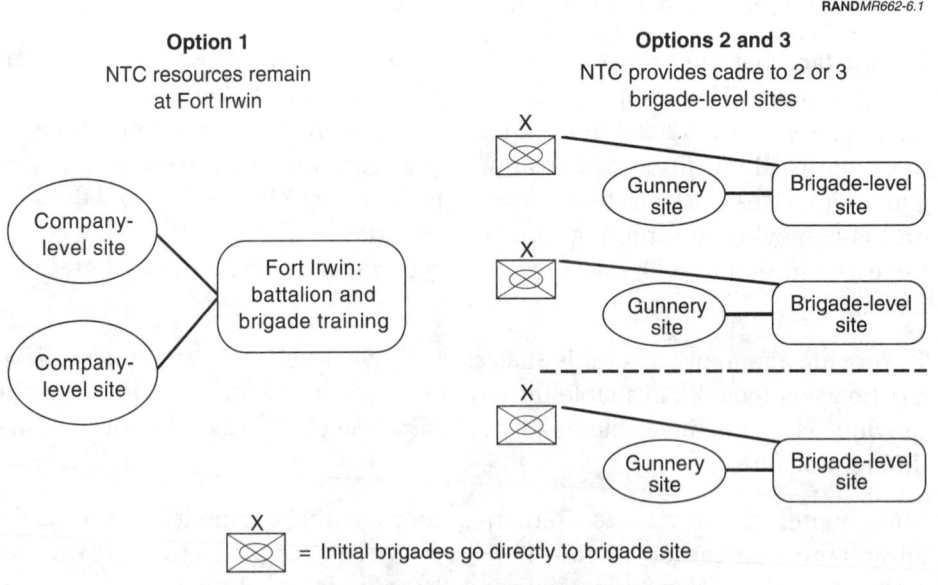

RAND*MR662-6.1*

Figure 6.1—Options for Executing Training Model

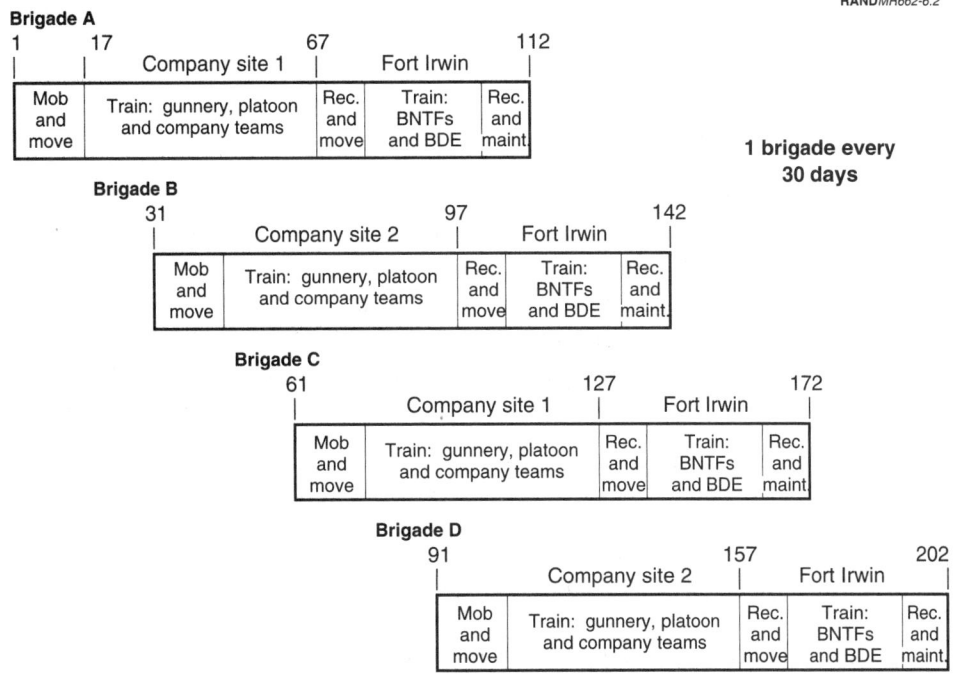

Figure 6.2—Two Company-Level Sites Feed Fort Irwin

company-level training site every 60 days. This scheme results in a flow of a new brigade arriving at the NTC every 30 days. The staggered arrivals allow one brigade to complete training at the company training site before the second brigade shows up, and they provide sufficient delay between brigades arriving at the NTC.[1]

The company training sites need to be relatively capable installations. They must have the full complement of gunnery ranges and sufficient maneuver space to conduct nine simultaneous company lanes. Any of the sites identified in Chapter Three other than Fort Irwin would be adequate.

The company training sites require the same number of training personnel that are required to execute the basic training model. We have already identified enough training personnel for three brigade sites, so the training personnel are available.

---

[1]The staggered start also allows NTC trainers to assist in training event and trainer preparation at the company-level training sites. This could be an important consideration for developing effective company lanes and preliminary command-and-control training.

The OPFOR at the company training sites is provided by the ARNG. As described in Chapter Five, the OPFOR requirement at a company-level training site is less than a full brigade training site. Additionally, OPFOR C2 and quality is less of a concern because of the reduced OPFOR echelon and the more structured nature of platoon and company lanes.

Fort Irwin, with some augmentation, has sufficient trainers and training support personnel to support higher-echelon training because gunnery qualification takes place at the company-level sites. The total number is slightly less than the typical brigade training site described in Chapter Four. Because brigades with three battalions are being trained, the NTC OPFOR would have to be augmented by about 1,000 personnel, as we discussed in Chapter Five.

Option 2: Two Brigade Sites

Under this alternative, two brigade-level training sites are established and are supported by two gunnery sites for the follow-on brigades. The resources of Fort Irwin—the O/Cs and the OPFOR—divide between Fort Irwin and one other brigade-level site (e.g., Fort Carson or Fort Bliss). The NTC cadre at each site provides expertise in higher-echelon training and OPFOR command and control.

The first two brigades to be mobilized report directly to the brigade sites and conduct all of their training there. To improve the rate at which ARNG brigades can be trained, subsequent brigades report first to a gunnery training site before moving on to the brigade site. Thus, while the initial brigades are completing their training at the brigade sites, subsequent brigades mobilize and move to the gunnery sites to commence training. Figure 6.3 depicts this concept as well as the option that would utilize three brigade sites (the C and F brigades in the diagram). For two brigade sites, four installations are required (two gunnery sites and two brigade sites).

At gunnery sites the brigades accomplish essentially the same training objectives as those shown for days 13 through 50 in the training model (refer to Figure 2.1 and Appendix A), with the exception of collective platoon training, which is accomplished at the brigade-level training site. Activities include individual and crew-served weapons qualification, common task training (CTT) validation, and MILES training for all units. Maneuver battalions execute crew gunnery qualification through Tank/BFV Table XII, infantry squad live fire, Scout Table XI, mortar subcaliber training, and crew and platoon battle drill training. Also accomplished is field artillery training through Artillery Table 8 (battery qualification), engineer training through platoon lanes and STXs, and Stinger gunnery training. Training aimed at giving the organizations a basic ability to sustain operations from a tactical configuration, and preliminary

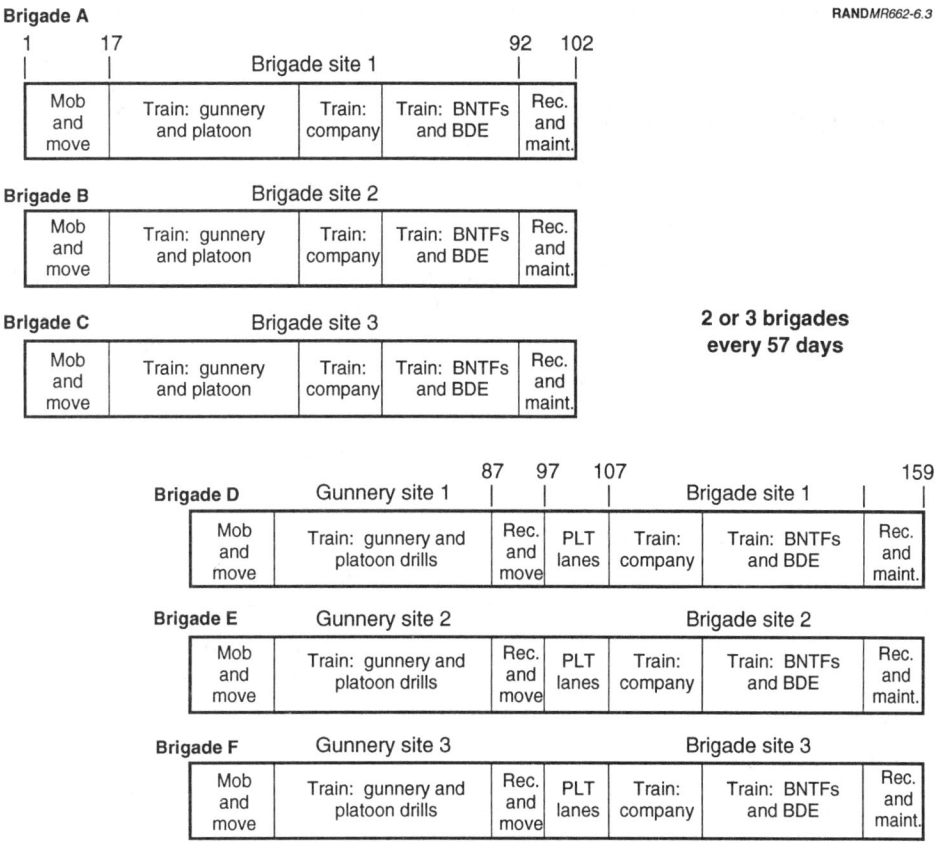

Figure 6.3—Two or Three Brigade-Level Sites

command and staff training consisting of individual and sectional functional training, orders drills, map exercise (MAPEX), and command post exercise (CPX) also take place at these sites.

Under this option the second set of brigades can have up to 75 training days at the gunnery training site, depending on when the brigade was mobilized. This is approximately 37 more days than allocated for similar training activities in the basic postmobilization training model at a single training site. The increase in available training time has the potential to reduce the requirement for training personnel at these sites, because far less use of parallel training is required.

The training activities at the brigade training sites for the first set of brigades include all activities described in the training model after initial preparations at the mobilization station. For the second set of brigades (those that go to a gunnery training site), this includes final command-and-control exercises, platoon through brigade collective training, plus recovery and maintenance services. Referring back to Figure 6.3, we estimate that platoon lanes would take about 10 days for the maneuver battalions, and it would take a total of 62 days to accomplish platoon through brigade maneuver training. During this 10-day period, the field artillery battalion would complete Tables IX through XI plus a battalion external evaluation (EXEVAL), and the engineer battalion would conduct company-level STXs.

The gunnery training sites do not require the maneuver space that the brigade sites do because only platoon drills will be executed. However, full gunnery range capability is required. In addition to the five primary brigade sites described in Chapter Three, a number of other installations could serve as gunnery sites, including Fort Stewart and Fort Riley. These sites would require about half the garrison support as the brigade sites because there is no requirement for an OPFOR.

In Chapter Four we identified sufficient active component training personnel to support three brigade training sites. Under this option, only two brigade sites would be fully manned. The remainder could be divided between the two gunnery sites. By extending the schedule somewhat at the gunnery sites so fewer gunnery ranges operate simultaneously, fewer training personnel are required at each site than would be needed at either a brigade- or company-level training site.

With a moderate schedule extension, sufficient training personnel should be available for the two gunnery sites by dividing the remaining available training personnel (about 500 per site), using the additional RTD (48), adding one more RTB headquarters (38) for training command and control, and providing the active component trainers from an additional BCST brigade for staff and simulation training (23).[2] Moreover, these personnel could be augmented by using TCEs, BCEs, and gunnery range support personnel available from the two brigade-level training sites once gunnery qualification is completed for the initial brigades. The brigades at the gunnery sites do not have to move to the brigade site until 87 days after the first brigades are mobilized (see Figure 6.3), which should allow sufficient time for the gunnery program to be completed effectively.[3]

---

[2]Assuming that the additional RTB headquarters and BCST do not have CFP or ARNG light brigade training commitments. See the discussion in Chapter Four of active component personnel available but not used, in particular Table 4.2.

[3]In the training model described in Chapter Two, about 46 TCEs and about 32 BCEs become available at Day 33, and all the gunnery range support personnel become available at Day 50 from each brigade training site as gunnery qualification is completed. These personnel would be moved from the brigade sites to the gunnery sites as needed. See Appendix A for a detailed schedule and Appendix C for detailed training personnel requirements at brigade sites.

Option 3: Three Brigade Sites

This option would require three brigade-level training sites and three gunnery sites. The organization at the three brigade sites is similar to that of the two brigade sites in Option 2, except that the NTC resources—the O/C-Ts and the OPFOR—provide cadre at Fort Irwin and two other brigade-level sites. The flow of training is as shown in Figure 6.3 and is the same as Option 2 except that there are three of each type of site instead of two, so six installations are involved. Like Option 2, each gunnery site would require garrison augmentation of about half that required for a brigade training site.

We have identified sufficient training personnel for three brigade sites *but not for the additional three gunnery sites*. Unlike the two-brigade training site alternative, in which the remaining training personnel could be divided between the two gunnery sites, the only active component personnel available for the gunnery sites would be the RTDs with each of the three brigades (144) and those AC personnel who support the RC and were not allocated to the three brigade sites.[4] To support the gunnery sites, the latter would include three unallocated RTB headquarters (114) to provide training command and control, active component trainers from three BCST brigades (69) to provide the preliminary battalion and brigade staff training, and personnel from three RTB engineer (33) and artillery training battalions (39).[5] As was the case in Option 2, additional TCEs, BCEs, and gunnery range support personnel could be made available from the brigade training sites when gunnery qualification of the first brigades is complete.

If the gunnery site schedule is extended to reduce the training personnel required and the active component trainers identified above can be made available, the brigades going to the gunnery sites should be able to accomplish the required gunnery site training with some self-help.[6] A primary concern with this alternative is the availability of the additional training personnel who may be needed to support other higher-priority training missions, e.g., training the units from the contingency force pool (CFP) or light ARNG brigades. If they are

---

[4]Again, refer to the discussion of personnel available but not used in Chapter Four and Table 4.2.

[5]Only six RTB headquarters and five BCST brigades are potentially available. This option would require all but one of the RTB headquarters and BCST brigades. Also, it could easily use all of the unallocated personnel remaining in the RTB engineer and artillery training battalions.

[6]We have discussed conduct of the gunnery site training events with Army leaders experienced with current training practices in enhanced brigades, as this training is similar in some ways to that conducted during AT. All believe that given the longer training time available at the gunnery sites under Options 2 and 3, the second and subsequent brigades with their RTD, minimal active component augmentation, plus the 65 TCE and BCEs from the brigade site, could effectively provide the required training.

not available for this mission, they would have to be provided from other sources.

ASSESSMENT OF OPTIONS

In assessing the options, we established criteria for each of them in five major areas: trainers, OPFOR, sites, the amount of additional RC mobilization that would have to occur, and force-generation rates. Table 6.1 summarizes the results of our assessment, and subsequent pages elaborate on the evaluations.

Training Personnel

This category includes trainers, training managers, and training support personnel.

Quality. The least risk to training quality occurs when trainers are used in roles that derive the most benefit from their comparative strengths. We identified sources for trainers and training managers that would provide the most experienced trainers. However, within this group of high-quality trainers, different groups have different comparative strengths. For example, the trainers in the NTC Operations Group at Fort Irwin are the most experienced in battalion task

Table 6.1

Assessment of Options

Criterion	Option 1	Option 2	Option 3
Training personnel			
Quality	Best match	Good	Cadre too few
Number	2,660	2,932	3,222
OPFOR			
C2 and quality	Best match	Adequate	C2 a problem
Number ARNG OPFOR	8,270	6,264	11,372
VISMOD/MILES	Best	None at site 2	None at sites 2 and 3
Sites			
Instrumentation	Best	None at site 2[a]	None at sites 2 and 3
Live fire	Best—BN level	Company	Company
RC mobilization	14,198	12,355	20,067
Force generation	Worst	Worse than #3	Best
1st brigade in:	112 days	102 days	102 days
3 brigades in:	172	159 (4)	102
6 brigades in:	262	226	159

[a]The Army is evaluating a portable battlefield instrumentation system called PRIME, but it will only support company-level exercises.

force and brigade maneuver and battalion-level live fire. The peacetime roles of the trainers in the RTBs and RTDs stress platoon-level exercises and crew gunnery qualification while occasionally supporting company-level lanes. To the extent that the site options employ these groups in roles outside their comparative advantage, the risk to the quality of training increases.

The best match between training strengths and sites occurs in Option 1, in which the NTC Operations Group conducts battalion task force and brigade-level training, and RTDs and RTBs cadre operate the company-level training sites. The Operations Group at the NTC is a group of highly trained observer/controllers. They operate as organizational teams (e.g., brigade, task force, artillery, forward support battalion) and accompany units training battalion task force and brigade force-on-force and live fire at Fort Irwin. The observer/controllers are, for the most part, experienced in the position and functional area they are training, and they go through an additional training program once selected for assignment to the NTC. They also gain a great deal of experience while at the NTC, and thus they can provide units with seasoned advice about the specific functional area. The ten organizational teams vary in size from about 40 to 70 people.

The RTDs and RTBs have their most current experience in training ARNG brigades. In peacetime, their expertise is in gunnery qualification, platoon, and occasionally company lane training. They have staff to cover all functional areas at this level but not to provide battalion and brigade staff or task force maneuver training. They too go through additional training when assigned to an RTD or RTB.

Option 1 also requires the least reorganization. This attribute both facilitates and potentially reduces the cost of the peacetime planning and preparation required to execute postmobilization training. The Operations Group at the NTC has a team that plans the scenarios for each active component brigade rotation conducted at Fort Irwin. Planning and preparation for higher-echelon postmobilization training of the ARNG brigades for the most part would not differ from planning any other rotation. Similarly, the RTDs and RTBs have significant experience in planning for and managing gunnery qualification and lane training for the ARNG, since they go though a similar process in planning Annual Training for the brigades in peacetime. Planning and preparation are key to the timely execution of the training model.

Option 2 would also provide good training. It would require more reorganization, and the planning and preparation for the brigade-level training sites would be more difficult than in Option 1. In the course of this study, we interviewed either the senior trainer or operations officer of each of the teams at the NTC, as well as two commanders of the NTC Operations Group. The consensus was that the Operations Group could be divided in half and provide good training at the battalion task force and brigade level at two brigade training sites, but they clearly preferred Option 1.

With regard to Option 3, the Operations Group felt that reorganization to cover three brigade training sites would spread their expertise too thin and disrupt team synergism. They believed this would affect the quality of training, especially at the brigade and battalion staff levels where they are the primary source of expertise in field training at these echelons. The reorganization, planning, and preparation would also be more extensive and more difficult. Moreover, with the potential difficulty in providing active component trainers to the gunnery sites, the training quality at those sites might not be on a par with the other options.

Number. Option 1 requires the fewest trainers, 2,660, because Fort Irwin would not have to support a gunnery qualification and lower-echelon training program. As a result, it would require 163 fewer personnel than a typical brigade-level training site. Option 2 would require 2,823 trainers, plus an RTD (48), one RTB headquarters (38), and the active component trainers from one BCST brigade (23) for the fourth site, to yield a total of 2,932. Under Option 3, 2,823 trainers would be required for the three brigade-level training sites plus those needed at the gunnery sites. The gunnery sites require three additional RTDs, RTB headquarters, and BCSTs, plus about 33 RTB engineer trainers (11 per site) and 39 artillery trainers (13 per site). If CFP and ARNG light infantry brigade training personnel needs allow these additional active component trainers to be made available, a total of 3,222 would be required for Option 3. As we mentioned earlier, if these additional trainers cannot be provided from the AC personnel who support the RC, they would have to be provided from other sources. Option 3 probably has the highest risk with respect to providing adequate numbers of trainers.

OPFOR

As we discussed in Chapter Five, a high-quality OPFOR is essential in executing the training model. OPFOR mistakes can adversely affect quality of training and the time it takes to train the unit. Each option will require ARNG OPFOR in addition to the full-time OPFOR at the NTC, the 11th ACR. A key criterion in comparing the options is the echelon at which the Guard OPFOR units will have to operate and exercise command and control: MRB and MRR command during battalion task force and brigade force-on-force "free play" are the most difficult and require the highest-level synchronization skills. Similarly, the echelon affects the level of training required in peacetime.

C2 and quality. Under Option 1, the 11th ACR augmented by an ARNG mechanized infantry battalion and engineers provides the OPFOR for the battalion task force– and brigade-level training at Fort Irwin. ARNG OPFOR brigade-sized units support platoon and company lanes at the company-level training sites. Lanes are highly structured and OPFOR synchronization requirements are minimal when compared to higher-echelon force-on-force battles. The highest echelon required is two motorized rifle companies needed for the BLUEFOR defensive lanes. Peacetime training could focus on platoon and company proficiency. Moreover, under this option, the 11th ACR has time (about 38 days after mobilization and movement) to work with ARNG OPFOR at the company training sites to prepare them for each of the lanes the ARNG units would have to support. We believe that this option potentially provides the best match of OPFOR skills with the level of training to be supported.

Option 2 requires the 11th ACR to provide an OPFOR cadre to Fort Irwin and one other brigade-level training site. The ARNG provides an augmented brigade to round out the OPFOR at two sites. As we suggested in Chapter Five, the 11th ACR operates in a manner that allows it to assume command and control of the MRR and MRBs at two sites. Peacetime training for the ARNG OPFOR would focus on platoon- and company-level proficiency, but more emphasis would have to be placed on operations in the "free play" environment of higher-echelon force-on-force training. Again, the 11th ACR has time to provide additional training after mobilization. In our discussions with the 11th ACR leaders, they indicated that they could make this option work given a dedicated ARNG OPFOR and focused OPFOR training in peacetime, preferably at Fort Irwin.

The 11th ACR provides a cadre to three sites under Option 3. It could not provide the command and control for all of the MRRs and MRBs. The ARNG OPFOR leadership would have to command at least one MRR (a brigade-sized unit) and four MRBs. Figure 6.4 compares C2 OPFOR requirements in Options 2 and 3. The unshaded units in Option 3 are those the ARNG will have to command and control.

Because we require an OPFOR that is at least as proficient as the BLUEFOR unit being trained, this option presents a serious command-and-control problem. The peacetime training requirement would escalate significantly, and the 11th ACR leaders thought that it would be difficult to support the training of two ARNG brigades. Moreover, a smaller 11th ACR cadre at each site would have only 38 days after mobilization to bring the Guard units up to a much higher level of proficiency: brigade-level maneuver. We believe that from an OPFOR perspective, this is the highest-risk option.

RAND*MR662-6.4*

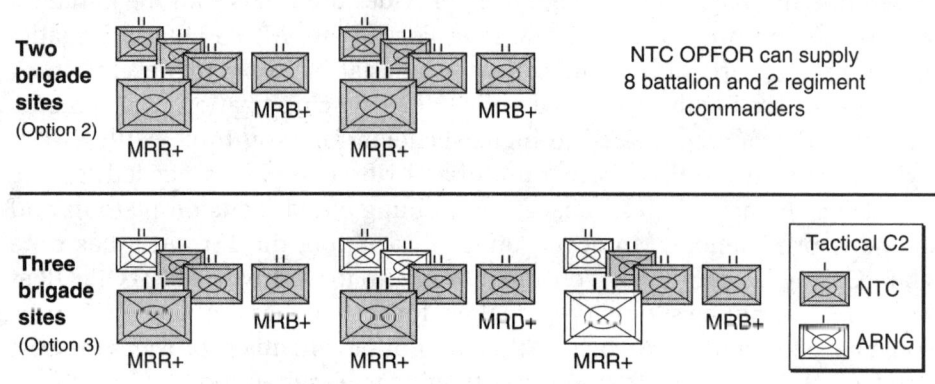

MRR = Motorized rifle regiment
MRB = Motorized rifle battalion
"+" means reinforced with tank units

Figure 6.4—OPFOR Command and Control Options

Number. Option 3 takes the largest number, 11,372, of ARNG OPFOR to support three brigade-level training sites. Option 2 requires the least, 6,264, because only the two brigade sites require OPFOR. Option 1 needs OPFOR support at three sites, but the company-level sites require fewer OPFOR than the brigade sites (refer to Table 5.1). The total requirement for this option is 8,270.

VISMOD/MILES. The 11th ACR at Fort Irwin is organized to replicate a reinforced motorized rifle regiment. It contains three maneuver elements, motorized rifle battalions, each containing about 13 tanks and 40 infantry fighting vehicles (BMPs). It also has a regimental reconnaissance element, artillery, and other regimental assets. Its main combat vehicle, the M551 Sheridan, has been visually modified to resemble T80 and T72 tanks, and the infantry fighting vehicles look like Russian BMPs. In addition, the 11th ACR's equipment has special MILES gear installed that mimics the performance of the weapon systems being replicated. Other support vehicles have been similarly modified.

Unfortunately, the NTC OPFOR equipment is unique to Fort Irwin and cannot be practically supported elsewhere. The OPFOR at other postmobilization training sites will have to use BLUEFOR equipment as is, because the cost of modifying that equipment would be prohibitive. Because this special equipment has the most value during higher-echelon force-on-force training, Option 1 is favored under this criterion.

Sites

In Chapter Three we identified five installations that can completely support the training model described in Chapter Two, although they differ somewhat in capability and availability. Sites are available that can support all three options, but Option 1 has a somewhat better match in terms of site capabilities, as discussed below.

Fort Irwin has unique capabilities not found at other installations that should figure into any postmobilization training plan. One is the sophisticated instrumented battlefield installed there, and the other is the live-fire area, which can support up to a brigade-level CALFEX.

Instrumentation. The instrumentation of the battlefield records the position of every combat vehicle over time as well as when those vehicles fire. In about one-quarter of the cases, the instrumentation can determine which vehicle is responsible for hitting another. In all cases, vehicle kills are recorded. The system also supports the play of artillery and other supporting systems. Video recorders situated at key areas supplement this instrumentation. This equipment allows the battle to be replayed on a large screen so that the tactics of both sides can be analyzed in detail during after-action reviews. This support enhances the training value of any force-on-force event, especially those at the battalion task force and brigade level. The instruments are permanently emplaced at Fort Irwin and could not be easily moved to another installation to support reserve component training. Although Fort Irwin would be used for each option, other sites under Options 2 and 3 would not benefit from such a capability.

CALFEX. Because of the live-fire capability at Fort Irwin, under Option 1, all ARNG brigades would be able to execute battalion-level live-fire exercises with true synchronization of other support at brigade level. Other locations would be limited to company-level CALFEXs with limited synchronization opportunities.

RC Mobilization

In the preceding chapters, we identified the requirement to mobilize reserve component personnel to support the postmobilization training of heavy ARNG brigades. This requirement includes some trainers, training support personnel, OPFOR, and units to augment garrison support. The latter requirement could vary significantly depending on the installation and the availability of additional civilian personnel or contractor support. Table 6.1 summarizes the mobilization requirement, and Table 6.2 shows the requirement by category.

Table 6.2

Reserve Component Mobilization Requirements

	Option		
	1 Brigade + 2 Company	2 Brigade + 2 Gunnery	3 Brigade + 3 Gunnery
Training	723	886	886
OPFOR	8,270	6,264	11,372
Garrison augmentation	5,205	5,205	7,809
Total	14,198	12,355	20,067

Under the worst-case assumption, that all garrison support augmentation would be provided by reserve component units, Option 3—with three gunnery sites and three brigade-level training sites—has the largest mobilization requirement, 20,067. Option 2 requires the fewest to be mobilized, 12,355, because only the two brigade-level sites require OPFOR. Option 1 requires slightly more, 14,198, because of the additional OPFOR requirement at three sites, although each company-level training site requires fewer OPFOR than a brigade site.

Force Generation

Thus far, the discussion has revolved around the time it takes a single unit to prepare and the requirements to support the different alternatives. However, also of interest is the rate at which the alternatives can generate trained brigades. The remainder of this chapter describes the factors that affect the rate at which brigades can be generated and illustrates the rate each alternative can support.

Factors affecting rate. If basic assumptions are met, three factors affect the rate at which brigades will be available in theater: the number of training sites, the missions the RC brigades will perform, and the availability of additional brigade sets of equipment.

Number of training sites. The number of sites affects the generation rate significantly. Table 6.1 shows the rate at which brigades can be trained for a combat mission and be ready to deploy for one brigade site with two company-level feeder sites. This option can generate a trained brigade every 30 days after the first. The first brigade would be ready to deploy at M+112, and six could be made available by M+262. Option 1 has the slowest force-generation rate.

As shown in Table 6.1, Options 2 and 3 have faster generation rates, with Option 3 generating the most in the shortest period. Under these options, two or three brigades can be trained and ready to deploy in 102 days. Six brigades can be made ready under Option 3 in 159 days and under Option 2 in 226 days.

Missions of the enhanced brigades. The missions the brigades train for can also influence the generation rate. Up to now our discussions have focused on the time it takes to train a brigade to reinforce or augment an active unit and perform full tactical operations in a combat theater, the most challenging mission. But reserve brigades could fulfill other missions that would require less extensive training and thus be made available sooner. Two additional missions were identified for the ARNG brigades by the Enhanced Brigade Task Force:[7] backfill and rotation. The former mission requires the reserve unit to replace a Europe-based active unit providing a forward presence that deploys to a major regional contingency. In the latter mission, the ARNG unit replaces an active component unit in an area of operations where a security force or stabilizing presence is required. Both of these missions require the reserve brigade to have credible combat skills, but they would not be as demanding as performing full tactical operations in combat.

The reduced demands of these other missions could significantly reduce the training required before deployment. The training requirements depend heavily on the METT-T and the training strategy employed. For example, if ARNG brigades were to be sent to Europe to replace active component forward-presence forces, the CONUS training requirement would depend heavily on any likely threat they might encounter. If the threat was minimal, they could be mobilized and flown to Europe to fall in on prepositioned equipment already there. They could then proceed with training at Grafenwöhr and the CMTC.

For some security missions, the brigades may require only specialized field training through company level with the capability to provide command and control as well as support at the level organized. Under these circumstances, they could be ready to deploy in as few as 60 days.[8] Other such missions that have the potential to escalate quickly to a full-combat-capability requirement could take the full 102 days of preparation.

Additional sets of equipment. The discussion thus far has assumed that brigades would train on their own equipment and, after recovery from training, would move it to the port and transport it to the theater of operations. This

---

[7]Headquarters Department of the Army (1994).

[8]They would execute gunnery qualification through company lane training with a focus on the METT-T of the security mission, followed by ten days of recovery and maintenance before deploying.

means trained brigades arrive in theater 22 to 30 days after the completion of training, depending on which theater they deploy to, the first brigade(s) arriving on M+124 or M+132. Brigades can be ready to operate in theater faster if additional sets of equipment can be prepared for shipment and placed in transit while the brigades complete training. The first brigades could then be ready to operate at around M+108.

In a two-MRC scenario, a number of active component units are slated to fall in on prepositioned equipment in theater, leaving their own equipment at home. If that equipment could be prepared and made available to ship in parallel with ARNG brigade training, the time required to generate trained ARNG brigades in theater would improve significantly.[9]

Do the force-generation rates meet requirements? We do not assess the capability of the options to meet requirements, because they are classified. We note that meeting requirements is more than a matter of the brigades' training time. Also at issue is the availability of transportation to move the units to theater. Under any scenario that would require the mobilization of these brigades, all of the active component TOE units and their support elements (including the reserve component units in the CFP) would be deploying to theater. Transportation could well be the constraint.

Other scenarios would not depend as heavily on transportation. For example, if the ARNG brigades were backfilling units in Europe, they might be able to deploy personnel by air and use either POMCUS equipment or the equipment of the deploying unit (which could use prepositioned stocks).

SUMMARY OF OPTIONS FOR EXECUTING THE MODEL

Policymakers have three mobilization options, ranging from one to three brigade-level sites. As the number of sites increases, the force-generation rate increases, but the increase in difficulty of execution is exponential, not linear. As more sites are activated, more organizational accommodations have to be made, and the pool of critical expertise spreads more thinly. The planning and preparation required before mobilization will also increase significantly. All of this increases the potential for something to go wrong. As we have indicated earlier, the model allows for some slippage, but any significant problem will either cause the schedule to slip or result in a lower level of training achievement.

The number of reserve personnel mobilized to support the different options ranges from about 12,000 to over 20,000. Our assessment of the options across

---

[9]Some believe that the equipment left behind would not be available because the CINCs would want it shipped to be used as war reserve materiel.

the dimension of trainers, OPFOR, sites, and force generation suggests that siting the training at Fort Irwin and two company-level sites provides the best approach in terms of training quality and risk. However, it generates forces at the slowest rate.

Choosing among the options requires a tradeoff among risk, resources, and force generation. The available assets will support any of the three options, but the three-brigade option poses the greatest risk (training quality, timeline, and availability of trainers), requires the most resources, but delivers the most brigades in the shortest time.

Should the mission of the enhanced brigade be modified, the training requirements could be relaxed, and thus the brigades could be available sooner. However, the resources required would not diminish significantly if training to company-level proficiency remained the goal and the same training timeline was to be maintained.

Chapter Seven

CONCLUSIONS AND IMPLICATIONS

This chapter summarizes the major conclusions of our analysis and examines some of their implications. The analysis assumes that the Army's objective is to produce fully trained ARNG maneuver brigades, ready to enter combat upon arrival in the overseas theater. These brigades are to be prepared as quickly as feasible, given the foreseeable resource constraints. We examined options that include running one, two, or three training sites simultaneously.

CONCLUSIONS

Our major conclusions are as follows:

Enough resources exist to support the training model at up to three brigade-level sites.

Of the options we identified as feasible, *operating three brigade-level sites* (with three preliminary gunnery sites) is the most demanding but also the fastest. It imposes a substantial resource bill, including about 1,900 active component trainers and almost 20,000 reserve component support personnel who must be mobilized. It also poses some risks to the quality of training and the training timetable. For example, to support three sites, the AC trainers must be split so that the group at one site lacks substantial experience in training battalion and brigade echelons. Furthermore, the challenge of developing a proficient OPFOR becomes greater when a third site is activated. However, the three-site option has the fastest force-generation potential: If all goes well, it would produce three brigades in 102 days and six brigades in 159 days.

The decision on how many training sites to operate will hinge on tradeoffs among risk, resources, and force-generation rates.

Operating two brigade-level sites (with two preliminary gunnery sites) poses less risk. The existing set of AC trainers can readily be split between two sites and still provide good training oversight at higher echelons. Similarly, it is more likely that the Army can prepare a capable OPFOR for two sites than for three. But the two-site option is somewhat slower in generating forces. It produces two trained brigades in 102 days, four brigades in 159 days, and six brigades in 226 days.

Operating one site is another alternative, with still less risk. Under this option, Fort Irwin would be the single site for battalion- and brigade-level training, and two other sites would accomplish gunnery through company-level training. This option provides the best match of trainer skills and OPFOR requirements and thus the least risk in terms of training quality and meeting timelines. However, it requires slightly more training resources than two brigade training sites, and it results in the slowest force-generation rate. The first brigade would complete training in 112 days, and one additional brigade would be produced every 30 days after that (at 142 days, 172 days, and so forth).

The Army's ability to generate forces and to meet the training timetable rests on a number of significant assumptions, some of which may be optimistic.

- Mission

  — The goal is a unit trained to enter combat upon arrival in theater.

- ARNG brigade readiness

  — The enhanced brigades will have a training status equal to the better brigades observed during Annual Training in 1992 and 1993.

  — The brigades can reach a C-1 status in both personnel and equipment by M+18.

  — Their maintenance skills will allow them to sustain their equipment through the 80 intensive days of field training.

- Logistics support

  — Spare parts and training ammunition will be available at each site to support the high operational tempo required by the training model.

- OPFOR readiness

  — ARNG OPFOR units will be available at the training site, ready to support platoon lanes and enter their own higher-echelon collective training by M+18.

  — The OPFOR will be at a sufficient level of training readiness to be able to complete most of their higher-echelon collective training by M+50.

- Training organization

  — A training organization will be in place with necessary equipment, planning, and preparation completed to commence training by M+12.

IMPLICATIONS

These conclusions suggest a number of implications. Some of these pertain to the ARNG and others to the Army at large.

Implications for National Guard Units

ARNG brigade mission. The focus of our analysis has been on the time and resources it takes to train a brigade to reinforce or augment an active unit in a combat theater, the most challenging mission. But reserve brigades could fulfill other missions that would require less extensive training and thus make them available sooner. Two additional missions were identified for the ARNG brigades by the Enhanced Brigade Task Force:[1] backfill and rotation.

We describe the characteristics of these missions in Chapter Six. The primary implication of changing the mission is that it will affect the training timeline. For the backfill and rotation mission, the time required for training would decrease substantially. However, the amount of resources required would not decline very much.

Enhanced brigade readiness. We have made some significant assumptions about the peacetime training achievements and personnel readiness of ARNG brigades as they enter postmobilization training. Currently, several of the enhanced brigades would have difficulty meeting the requirements implied by these assumptions. Their primary problems center on maintaining personnel readiness (a high rate of fill with MOS-qualified soldiers). These personnel shortfalls require the brigades to send many soldiers to school instead of attending Annual Training with their unit, thus impeding the unit's collective training. As a result, a number of brigades would find it difficult to meet the timelines described in the training model.

Peacetime OPFOR in the National Guard. To execute the training model for any of the alternatives described here, the Army would need an ARNG OPFOR to augment the 11th ACR (the professional OPFOR at the National Training Center). Because training will begin soon after mobilization, the OPFOR must

_____

[1]Headquarters Department of the Army (1994).

be prepared virtually upon mobilization. Thus, some ARNG units must focus on OPFOR duties as a primary peacetime mission.

Again, the more sites contemplated, the larger the number of ARNG units that must have a peacetime mission of practicing OPFOR skills. The three-site option could involve several thousand reserve component personnel. Under that option, the Army might also face added resource costs in training days and operational tempo costs to keep these units at the necessary level of proficiency. In addition, more active component personnel would be needed to serve as OPFOR reserve training detachments and to augment the 11th ACR for coordination of peacetime OPFOR training.

Postmobilization training of ARNG divisions. The training model prepares the enhanced brigades for combat as expeditiously as possible. That said, it still takes over three months to deliver three brigades to theater and six months to deliver six brigades. These units enjoy a higher level of equipment and training support in peacetime than the ARNG divisions and are presumably the most ready combat units among the reserve components. It would seem that ARNG divisions would not begin training until the enhanced brigades had completed their training. Moreover, the divisions would most likely take longer to reach deployment standards. As with the ARNG separate brigades, the training would depend on the METT-T of the theater to which the divisions would be deployed. Thus, any plans that involve ARNG combat divisions should assume that the better part of a year would pass after the first separate brigade has been mobilized before any divisions were ready to deploy.

Implications for Other Army Elements

Logistics support. The training model implies that each training site will require significant quantities of spare parts and training ammunition to sustain the high operating tempo of the brigade being trained, the OPFOR, and the trainers. Based on our discussions with personnel in the Department of the Army and Forces Command, it appears that this requirement has not yet been formally identified nor sourced. During mobilization, requisitions for spare parts at brigade training sites will compete with requisitions from units that have already deployed or that are about to deploy to a combat theater. The ARNG units would surely have the lower priority.[2] As a result, the Army might be unable to sustain training operations at the brigade training sites, which could significantly lengthen the time to prepare the brigades.

---

[2]As would active component units that were also preparing to deploy during the same period.

Premobilization planning and preparation. None of the options described in this report can occur without substantial planning and preparation in peacetime. To begin training on M+12, the organizations involved would have to have done considerable advance planning and preparation, regardless of the number of sites operated. Even the one-brigade training site option requires establishing two company-level training sites. This means that planning and preparation has to occur in peacetime, at the installations that will be involved, by the organizations providing the training personnel, and by the units supplying the support forces. This additional peacetime activity would fall on top of current requirements. We have not attempted to quantify the scope of that activity. It is not trivial, however, and the more sites contemplated, the greater the peacetime planning and preparation required.

Peacetime costs of reserve component postmobilization training. Analysis of the resource requirements suggests that the enhanced brigade structure imposes some additional costs that should be taken into account. An acknowledged cost already being paid is the active support provided to reserve component training during peacetime. But the planning and preparation, the peacetime OPFOR, and garrison support requirements described above imply additional costs for resources that will primarily be used to support the postmobilization training of ARNG brigades. Furthermore, as we discussed above, additional funding may be required to ensure that adequate spares and training ammunition are available to support the intensive training program. The costs could be substantially curbed if the timelines of deploying the unit could be relaxed. More time could both shrink the number of installations required and eliminate the need for parallel training, which drives up the number of trainers needed and, hence, the costs.

DETAILED DESCRIPTION OF THE TRAINING MODEL

This appendix provides a detailed description of the training model described in Chapter Two. At the end of this appendix is a long table portraying the day-by-day schedule of the training events in our model for each element of the brigade. We will describe each horizontal block in this table, starting with brigade and battalion command and control, continuing with the maneuver battalions and then through each brigade slice element.

BRIGADE AND BATTALION COMMAND AND CONTROL

Overall Description

Command-and-control training progresses from individual through brigade-level force-on-force exercises at the end of the training period. Preliminary command-and-control training applies to all commanders and staffs in all companies, battalions, and the brigade headquarters to progress from individual and section to command and staff exercises to prepare them for collective training exercises.

The objective is for leaders to be prepared by the time collective training begins, so that collective training can focus on collective performance and training rather than on leader training. The preliminary C2 training and leader training exercises are scheduled to allow battalion and company commanders to observe and participate in training exercises with their subordinate units. Additionally, it is structured so that the brigade and battalion chain of command controls training activities and performs normal sustainment functions from a tactical configuration concurrently with conduct of the leader training. An example of participation would be for the company commander, first sergeant, and FIST to participate as a command group during the orders, preparation, and execution portions of platoon lane training and then to participate in the execution portions of the battalion MAPEX and CPX during late afternoons and evenings of the same day.

Another key aspect of preliminary C2 training is that we have allocated a full set of trainers for maximum leader training benefit during these exercises. Following the same example above, the trainers working with the company commander, company first sergeant, and executive officer would be using the platoon lanes as training events to ensure that company sustainment and command-and-control activities are being conducted to standard and that company, battalion, and brigade systems will be ready for the more demanding higher-echelon training. This also provides time for more experienced trainers to conduct one-on-one training with the less experienced brigade leaders. Below we describe the sequence of individual events.

- **Command and staff functional training.** Individual and staff section refresher training is conducted by command and staff trainers, ensuring knowledge of tactical and technical concepts and staff procedures and responsibilities.

- **Command post setup drills and command group orders drills.** Refresher training is conducted to ensure that the command post can operate and move and that commanders and staffs can produce a doctrinally correct, properly coordinated operations order (OPORD).

- **MAPEX and seminar.** Commanders and staff are given a mission and tactical situation, are walked through the development of an OPORD, and execute on a simulation. These are done as separate exercises for each battalion and for brigade. In addition to O/C-Ts cover down, this requires simulation and simulation support personnel. Brigades write OPORDs for battalion CPX.

- **CPX (command post exercise).** Units go to field location, set up, and develop and execute an OPORD (in a simulation). These exercises require the same support as MAPEX. They also require net control station personnel and staff to act as higher headquarters, monitor nets, and provide situation information and control.

- **C2 of subordinate unit training.** Under O/C-T cover down, commander and staff prepare OPORDs and perform doctrinal C2 and sustainment roles during subordinate unit exercises. Notice that the battalion CPX are completed in time for the full-time attention of the battalion commander and staff on this activity during company team lanes training.

- **Theater war planning.** Brigade staff begins to plan for wartime contingencies following guidance provided by the theater commander. During maneuver training events the command-and-control training continues. The events are conducted with structured leader training preceding each event. The amount of time provided increases with echelon. During platoon lanes it is part of a day. A day is allocated to company team lanes, and two days in battalion FTXs. During the battalion task force FTXs, there is a day available to plan and a day to conduct preparation, which can include a full-scale rehearsal or a command field exercise (CFX). Again, the model provides the time and trainer resources for leader training before collective unit exercises throughout the training period.

MANEUVER BATTALION TRAINING

Overall Description

Maneuver battalion training consists of several major blocks of training, and these vary somewhat depending upon the branch of the maneuver unit. For example, the infantry battalions schedule a period for a BFV TOW firing exercise, and the tank battalion does not. The sequence will also vary depending upon when a battalion has its Table V–XII gunnery scheduled. The major blocks of training are the following:

- Maintenance, individual training (CTT), crew/platoon stabilization, COFT
- Preliminary gunnery
- Individual/crew weapons
- Gunnery Tables V–XII
- Squad live fire
- Platoon battle drills
- Platoon-level situational training exercises
- Task force organization
- Company-level situational training exercises
- Battalion operations
- Brigade operations.

Maintenance, individual training (CTT), crew/platoon stabilization, COFT. During this period, refresher training is done on critical individual and crew tasks necessary for successful collective training such as CTT. Additionally, crew and organizational rosters are finalized. A key element is COFT exercises, to reach appropriate reticule aim groups. COFT training for weak crews continues until the crew is qualified.

Preliminary gunnery training is conducted in accordance with FMs 17-12 and 23-1. The brigade runs most of these events with the assistance and support of the active component (AC) trainers. This period includes

- TGST: tank gunnery skills test.

- BGST: Bradley gunnery skills test. Brigade trainers conduct the test after training and validation is conducted by AC trainers.

- TCPC: Tank crew proficiency course.

- BCPC: Bradley crew proficiency course. AC trainers set up TCPC and BCTC and provide TCE/BCE.

Individual/crew weapons. This training includes SAW, AT-4, Claymore, demolition, hand grenade, Dragon, M203 qualification, and 9mm and M16 qualification. It is conducted for each organization in the brigade.

Tank/Bradley/TOW Tables V–XII are conducted in accordance with FMs 17-12 and 23-1. To ensure completion of these tables to prescribed standards, these ranges are set up and the training supervised by the AC trainers. Also, range support personnel from outside the brigades are provided from AC or ARNG sources.

Squad live fire. The squad live fire runs concurrently with tank/BFV crew training and qualification in preparation for qualification on Bradley Table XII. Again, these exercises are conducted by AC trainers and have external range support personnel.

Platoon battle drills. Platoons and sections practice mounted and dismounted platoon drills and critical tasks including executing and changing formations, performing movement techniques, reaction to contact, react to ambush, secure at halt, air attack, react to indirect fire, and mounting and dismounting the Bradley. These are conducted without OPFOR.

Platoon lanes. Platoon lanes are designed as segments of the battle rather than as complete operations. The objective is to train all the platoon-level tasks necessary for participation in higher-echelon exercises. Company commanders control, give orders to, and assist in the evaluation of their platoons. Platoons report to company, evacuate casualties, perform resupply, and call for fires, and they are evaluated on these functions, as are the company command groups, FIST, and company trains. Permanent company as well as platoon O/C-Ts participate in these exercises. Company O/C-Ts mentor the company commander, executive officer, FIST, and first sergeant to use these events as leader training in preparation for company lanes. Specialization occurs depending on the type of platoon: infantry platoons undertake "trench" lanes, while tank platoons undertake "breach" lanes. Lanes are as follows:

- **Hasty defense (HD).** Platoons defend against an OPFOR of two reinforced motorized rifle platoons. Tasks include perform tactical planning, prepare for combat, maintain operational security, sustain, occupy an assembly area, move tactically, react to contact, occupy a battle position, defend, employ fire support, perform surveillance, perform attack by fire, move tactically, perform passage of lines, defend, and consolidate and reorganize.

- **Trench.** This STX is for Bradley platoons only. Platoons attack an enemy trench line defended by an enemy squad. Tasks trained include prepare for combat, maintain operational security, sustain, occupy an assembly area, move tactically, assault, breach obstacle, clear a trench line, and consolidate and reorganize.

- **Attack by fire (AbF).** Platoons attack by fire an OPFOR of a dug-in reinforced motorized rifle or tank platoon. Tasks trained include perform tactical planning, produce a platoon fire plan, prepare for combat, maintain operational security, sustain, occupy an assembly area, move tactically, occupy a battle position and defend, employ fire support, perform surveillance, and perform attack by fire.

- **Hasty attack (HA).** Platoons attack an enemy outpost occupied by section (two armed reconnaissance vehicles, tanks, or BMPs). Tasks trained include perform tactical planning, prepare for combat, maintain operational security, sustain, occupy an assembly area, move tactically, react to indirect fire, react to contact, perform reconnaissance by fire, employ fire support, perform fire and movement, assault, and consolidate and reorganize.

- **Breach (B).** This STX is for tank platoons only. Platoons conduct a mechanical breach of a mine and wire obstacle covered by direct fire. Tasks trained include perform tactical planning, prepare for combat, maintain operational security, sustain, move tactically, take actions at an obstacle, and perform passage of lines.

Task force organization. Companies and maintenance elements are cross-attached between maneuver battalions to form company teams and battalion task forces. Almost all of the brigade slice training and battalion-level preliminary C2 exercises are completed by this time to allow the focus to shift toward actual command, control, and support of combat operations. During this period, company teams practice basic movement techniques, drills, and sustainment and support activities without an OPFOR.

Company STXs. Company teams perform lanes designed to train the tasks needed for successful participation in battalion task force–level exercises. As with platoon lanes, exercises are not complete operations but segments of the battle, and specialization occurs. These are also structured training events for the battalion commanders, staff, command posts, and support elements. Battalions control and give orders. Company teams report, evacuate casualties, conduct resupply, call for fires and perform other support through battalion, and they are evaluated on these functions. Lanes are as follows:

- **Counterreconnaissance (CR).** Company teams perform counterreconnaissance against OPFOR regimental reconnaissance elements. Tasks trained include prepare for combat, maintain operational security, perform logistics planning, perform resupply, perform maintenance activities, provide medical evacuation and care of casualties, occupy an assembly area, move tactically, conduct screen operation, employ fire support, withdraw and perform passage of lines.

- **Deliberate defense (DD).** Company teams defend against two reinforced motorized rifle companies. Tasks trained include prepare for combat, maintain operational security, perform logistics planning, perform resupply, perform maintenance activities, provide medical evacuation and care of casualties, occupy an assembly area, move tactically, establish an obstacle, defend, employ fire support, and consolidate/reorganize on an objective.

- **Reposition in the defense (REPO).** Company teams occupy a battle position and reposition to conduct an attack by fire on an OPFOR of two reinforced motorized rifle companies. Tasks trained include prepare for combat, maintain operational security, perform logistics planning, perform resupply, perform maintenance activities, provide medical evacuation and care of casualties, occupy an assembly area, move tactically, establish an obstacle, defend, employ fire support, perform actions on contact, and attack by fire.

- **Movement to contact, hasty attack (Mv/HA).** Company teams perform movement to contact and hasty attack against a reinforced motorized rifle platoon. Tasks trained include prepare for combat, maintain operational security, perform logistics planning, perform resupply, perform maintenance activities, provide medical evacuation and care of casualties, occupy an assembly area, move tactically, perform actions on contact, employ fire support, assault, defend, and consolidate/reorganize on an objective.

- **Advance guard, support by fire (AG/SF).** Company teams perform movement to contact and perform support by fire against a reinforced motorized rifle company. Tasks trained include prepare for combat, maintain operational security, perform logistics planning, perform resupply, perform maintenance activities, provide medical evacuation and care of casualties, occupy an assembly area, move tactically, perform guard activities, perform actions on contact, employ fire support, and perform support by fire.

- **Breach (B).** Tank-heavy company teams perform a mechanized breach of a complex obstacle. Tasks trained include prepare for combat, maintain operational security, perform logistics planning, perform resupply, perform maintenance activities, provide medical evacuation and care of casualties, occupy an assembly area, move tactically, perform passage of lines, employ fire support, and breach.

- **Assault a trench system (trench).** Infantry-heavy company teams or infantry companies assault a reinforced, entrenched motorized rifle platoon. Tasks trained include prepare for combat, maintain operational security, perform logistics planning, perform resupply, perform maintenance activities, provide medical evacuation and care of casualties, occupy an assembly area, move tactically, perform passage of lines, employ fire support, assault, and consolidate/reorganize on an objective.

- **Assault (AS).** Company teams perform a mounted assault against a reinforced hasty defending motorized rifle platoon. Tasks trained include prepare for combat, maintain operational security, perform logistics planning, perform resupply, perform maintenance activities, provide medical evacuation and care of casualties, occupy an assembly area, move tactically, employ fire support, perform passage of lines, assault, and consolidate/reorganize on an objective.

Battalion task force operations (BN OPS). Battalion task forces prepare for brigade exercise by conducting FTX on the missions of deliberate attack (DA), movement to contact (MTC), and defense in sector (DS). Three days are allocated to exercise to allow structured, walk-paced preparation with leader training, rehearsals, and CFX prior to run-paced execution. This time also allows retraining to take place as necessary to ensure the reaching of appropriate standards. The brigade commander and staff control, give orders, and receive reports. Brigade elements perform all the tactical, battlefield operation system and sustainment tasks associated with the missions listed in ARTEP MTP 71-2 and 71-3.

Company combined arms live-fire exercises (CALFEX). Company team CALFEXs are conducted at a post other than the NTC. At the NTC this would be the normal brigade live-fire portion of the rotation. During the CALFEX, each company team conducts live-fire movement to contact with a hasty attack and a defense in sector. The cavalry troop also conducts a CALFEX. Engineer, mortar, and scout platoons participate with company teams.

CAVALRY TROOP: SEQUENCE AND DESCRIPTION OF INDIVIDUAL EVENTS

Gunnery for the tank platoons and cavalry platoons is scheduled with the tank and mechanized infantry battalions. There are some differences in gunnery requirements between cavalry fighting vehicles (CFV) and Bradleys; the cavalry platoons do not require the dismounted squad live fire, and cavalry section qualification tables (Scout Tables IX and X) are fired rather than platoon Table XII. Platoon lanes for the tank platoons of the cavalry troop are similar to those of the tank platoons except that no breach task is included, and the conditions are modified to align more closely with the troop mission. The STX for the cavalry platoons reflect their security and reconnaissance missions with appropriate tasks from ARTEP-MTP 17-57-10. The lanes are performed over a two-day period to give adequate time to plan, prepare, and execute the lanes at a walk pace and over doctrinal frontages and depths. The troop's mortar section trains with the rest of the brigade's mortars and finishes in time to participate in the troop-level STX. The following lanes are included as cavalry platoon lanes:

- **Screen a stationary force (ScrSF).** Cavalry platoons each perform forward security for a notional defending force, detect and defeat OPFOR reconnaissance elements (individual or pair of vehicles), report required information, and perform a passage of lines.

- **Screen a moving force (ScrMF).** Cavalry platoons perform a flank screen for a notional moving force to detect and report various OPFOR activities.

- **Route reconnaissance (RtRec).** Cavalry platoons perform reconnaissance of a designated route and adjacent terrain to identify trafficability and enemy activity and any other required information.

- **Area reconnaissance (ArRec).** Cavalry platoons perform reconnaissance of a designated area, identifying and reporting enemy and other required information.

- **Zone reconnaissance (ZnRec).** Cavalry platoons perform reconnaissance of a designated zone to identify enemy activity and any other required information.

Troop lanes cover the same tasks as the cavalry platoon as outlined in ARTEP-MTP 17-487-30 but include larger areas of terrain and larger OPFOR elements. Because of the presence of tank platoons, the troop would be expected to fight and defeat small OPFOR elements in the course of these missions. Additional tasks at troop level are hasty defense (HD) and movement to contact/hasty attack (MV/HA).

CALFEX and battalion/brigade operations. The troop conducts two live-fire CALFEX on the missions of hasty attack (HA) and hasty defense (HD). The troop participates in battalion- and brigade-level training by performing doctrinal missions under brigade control.

SCOUT PLATOON: SEQUENCE AND DESCRIPTION OF INDIVIDUAL EVENTS

Gunnery. The gunnery training for the scout platoons employs the new HMMWV tables recently developed by the Armor School.

Scout lanes. Scout training is consolidated for the brigade and completed in time for the platoons to participate in company team lanes. The platoon STXs include the same basic tasks as those in the cavalry platoon and involve the same type of OPFOR elements, but because the scout platoons have less firepower, they do not attempt to defeat enemy forces during these STXs and work in closer proximity to notional friendly forces.

MORTAR PLATOON TRAINING: SEQUENCE AND DESCRIPTION OF INDIVIDUAL EVENTS

Mortar platoon and training for the cavalry troop's mortar section is consolidated at brigade level. This training is progressive and begins with individual training for gunners and FDC members, culminating with completion of gunner and FDC examinations. Training progresses first to crew/FDC section level, then to section, and finally to platoon-level STXs, where movement and reconnaissance, surveillance, and occupation of position (RSOP) tasks are trained. During these STXs there are requirements for small OPFOR elements to train self-protection and security functions. Live-fire training includes both subcaliber and full caliber firing. A platoon external evaluation is conducted as the last event before the platoons and sections begin participation in company team/troop–, battalion-, and brigade-level maneuver exercises and company team CALFEXs.

FIELD ARTILLERY BATTALION

Training for the elements of the field artillery battalion is progressive through battalion-level external evaluations (EXEVALs) and participation in brigade-level operations.

Howitzer batteries. Training for the firing batteries begins with training of individual crews and FDC sections on individual and crew skills. The batteries conduct the Artillery Tables (AT) for crews (AT 1 and 2), platoons (AT 3, 4, and 5), batteries (AT 6, 7, and 8), and battalions (AT 9, 10, and 11). The appropriate preliminary training precedes these tables.

FIST and COLT. Training starts with individual and team-level skills training to prepare for live-fire exercises and maneuver STX. Scheduling training for FIST is complex because the FIST must participate in numerous, concurrent training activities. They support their associated maneuver organizations during platoon and company team–level STXs by performing their fire support functions. The FIST chief also participates in preliminary command-and-control training events and exercises with the company team commander he supports. Additionally, FIST are key participants in mortar and field artillery live-fire exercises.

Battalion commander and staff. As with the maneuver battalions, the field artillery battalion commander and staff and maneuver battalion fire support officers and fire support sections participate in structured command-and-control exercises both for the battalion itself and in conjunction with the maneuver battalions and brigade. They also control the training activities and sustainment of their subordinate elements.

Headquarters and service battery. The training of the special platoons and sections (e.g., Meteorological and Counter Battery Radar) in the HHB consists of technical training and assessment performed by cover down elements. The training of the combat service support elements parallels that of like elements in the maneuver battalions.

ENGINEER BATTALION: SEQUENCE AND DESCRIPTION OF INDIVIDUAL EVENTS

CEV crew qualification and demolition training for combat engineer squad personnel requirements, as outlined in the Department of the Army Pamphlet 350-38, are conducted concurrently early in the training cycle.

Platoon lanes. During these lanes, appropriate elements of the assault and obstacle platoon, and Volcanoes from battalion, are attached to the platoons. Lanes include the following:

- **Route clearance (RtClr).** Platoons clear small point obstacles along a designated route.

- **Emplace point obstacle (Em PO).** Platoons establish craters reinforced with mines.

- **Reduce point obstacle (Rd PO).** Platoons reduce a point obstacle.

- **Build a battle position (Bld BP).** Platoons build a company defense position for a tank and mechanized infantry platoon.

- **Hasty defense (HD).** Platoons establish a platoon battle position and defend against a reinforced motorized rifle platoon.

- **Breach (B).** Platoons conduct a stealth breach of a wire mine obstacle.

- **Trench (T).** Platoons clear a trench line.

- **Move tactically (MvTac).** Platoons move along a route while maintaining its security and reacting to enemy contact, using direct and indirect fire.

Engineer company lanes. Engineer companies conduct lanes with normal attachments from the battalion. As with the maneuver company team lanes, the battalion conducts appropriate control and sustainment activities. Engineer company lanes include the following:

- **Movement and occupation of assembly area lane (Mv/AA).** Companies conduct a tactical road march, then occupy and defend an assembly area. An OPFOR platoon conducts an ambush and probes the assembly area.

- **Emplace a complex obstacle (EmCO).** Companies emplace a larger complex obstacle over a two-day exercise.

- **Combat route clearance (CbtRt).** Companies clear a battalion MSR that has numerous enemy and natural obstacles and small OPFOR security elements.

FORWARD SUPPORT BATTALION

Training for the companies in this battalion focuses on technical training for individuals and sections, functional (e.g., casualty evacuation and care) and common (e.g., move and occupy an assembly area) collective tasks described above for the support elements of the maneuver battalions. As with all elements, support of the brigades' elements from tactical configurations using appropriate battlefield (rather than administrative) methods would be performed. Training on active component systems (SIDPERS, TAMMS, TACTIS) could require special attention. The foward support battalion command and staff elements participate in the C2 training events described above.

AIR DEFENSE ARTILLERY BATTERY: SEQUENCE AND DESCRIPTION OF INDIVIDUAL EVENTS[1]

Air defense artillery battery training begins with a short period of crew drills and then goes to gunnery training. The BSFV platoons go through the standard BFV tables with the 1st Mechanized Infantry battalion. Concurrently with the BSFV tables, Stinger gunners go through gunner's training and a gunner's evaluation in accordance with FM 44-18-1 at a moving target simulator (MTS). If an MTS is not available at the training site, the gunners travel to an installation that has one and return to the training site. The BSFV squads conduct Tables IX A and X A after completion of BFV Table XII and successful completion of the Stinger gunner's exam.

Platoon drills and lanes are conducted to train basic organizational tasks and the air defense tasks of defending friendly attacking and defending units, support and command-and-control facilities, and, for the Stinger platoons, area defense missions. As with other lanes events, the commander performs full command-and-control roles and the battery conducts sustainment activities as structured training events. These are coordinated with other training events to allow coordination with actual friendly forces to the extent possible. Leader planning and air defense coordination are stressed during this period, and careful scheduling is necessary as the battery commander, executive officer, and platoon leaders also participate fully in brigade and battalion preliminary command-and-control training events.

During maneuver company team lanes training, BSFV platoons continue platoon-level training under maneuver battalion control, and the air defense artillery battery conducts battery-level training with the battery commander controlling the Stinger platoons across the brigade's operating and support areas.

During battalion and brigade task force operations, the air defense artillery battery and its platoons perform normal tactical missions.

BRIGADE OPERATIONS

The brigade performs a deliberate attack mission with all three maneuver battalion task forces and other brigade elements against a defending motorized rifle regiment over a three-day period.

---

[1]This outline of the air defense artillery battery training program is based on the current proposed organizational design of this battery. It has two BSFV platoons, two Stinger platoons, and a headquarters. This organization has not been fully fielded, and modification of this program may be needed.

RECOVERY, MAINTENANCE SERVICES, AND PREPARATION FOR DEPLOYMENT

The brigade prepares for final deployment. This portion is conducted as a structured training event, especially for the conduct of services that are seldom performed by National Guard crews and organizational maintenance personnel during peacetime. During this period any needed final crew qualifications are performed.

ACRONYM LIST

The following acronyms appear on the detailed schedule at the end of this appendix.

AbF	Attack by fire lane
AG/SF	Advance guard and support by fire lane
AR TM	Tank company team: 1 tank platoon and 2 mech infantry platoons
ArRec	Area reconnaissance lane
ARTEP EXEVAL	Army training and evaluation program external evaluation
AS	Assault lane
AS STX/LFX	Assault situational training exercise and live-fire exercise
AT1 AT11	Artillery tables 1–11
B	Breach lane
BCPC	Bradley crew proficiency course
BDE OPS	Brigade operations
BGST	Bradley crew gunnery skills test
Bld BP	Build a battle position lane
BN EXEVAL	Battalion external evaluation
BN HQ COs	Battalion headquarters companies
BN OPS	Battalion operations
Btry RSOP	Battery reconnaissance, survey and occupation of position
CAV TRP	Cavalry troop
CbtRt	Combat route clearance stx
CEV Gunnery and Demo	Combat engineer vehicle gunnery and demolition training
CFX	Command field exercise
CO CALFEX	Company combined arms live-fire exercise
CO/TMs	Company teams
COFT	Conduct of fire trainer
CP and Command Group Drills	Command post and command group drills
CPX (w/Sim)	Command post exercise (with simulation support)

CR	Counter reconnaissance lane
CSS-Combat Service Support	The sections in the battalion headquarters company that perform CSS functions, e.g., maintenance, medical, etc.
CTT	Common task training
DA	Deliberate attack operation
DD	Deliberate defense operation
Def STX/LFX	Defend situational training exercise and live-fire exercise
DS	Defend in sector operation
Em CO	Emplace complex obstacle
Em PO	Emplace point obstacles lane
EN BN	Engineer battalion
FA BN	Field artillery battalion
FA Btry & Bn LF	Field artillery battery and battalion live fire
FDC	Fire direction center
FIST	Fire support team
FOTS	Observer training system
FSB	Forward support battalion
FSO/FSS	Fire support officer/fire support section
GST	Gunners skills test
HA	Hasty attack lane
HD	Hasty defense lane
I VIII	Scout gunnery tables I–VIII
IIN TM	Infantry company team: 2 mechanized infantry platoons and 1 tank platoon
IN CO -	Infantry company: 3 mechanized infantry platoons
LFX	Live-fire exercise
LFX EXEVAL	Live-fire exercise external evaluation
M	Maintenance day
M/P	Maintenance and preparation day
M/Rtn	Maintenance or retraining day
MAPEX and Seminar (Sim Support)	Map exercise and seminar (simulations supported)
MILES	Multiple integrated laser engagement system training
Move and RSOP STX	Reconnaissance, survey and occupation of position situational training exercise
MTC	Movement to contact exercise
MTC STX/LFX	Movement to contact situational training exercise and live-fire exercise
Mv Tac	Move tactically lane
Mv/AA	Movement and occupation of assembly area lane
Mv/HA	Movement to contact and hasty attack lane
OW	Overwatch lane
P	Practice
Plt Prep	Platoon preparation

PMCS	Preventive maintenance checks and service
POM	Preparation for overseas movement
Rd CO	Reduce complex obstacle lane
Rd PO	Reduce point obstacle lane
Rd WO	Reduce wire obstacle
REPO	Reposition lane
Rt Clr	Route clearance lane
Rtn	Retraining
Rtn/M	Retrain/maintenance
Rtn/P	Retrain/practice
RtRec	Route reconnaissance lane
SbF	Support by fire lane
SCPC	Scout crew proficiency course
Scr	Screen lane
ScrMF	Screen a moving force lane
ScrSF	Screen a stationary force lane
STX	Situational training exercise
Sub Cal LFX	Subcaliber live-fire exercise
T	Trench clearing lane
TCPC	Tank crew proficiency course
TF	Task force
TGST	Tank crew gunnery skills test
TOW	TOW qualification table
Trench-Assault	Trench clearing lane
V–XII	Tank and BFV tables V to XII
ZnRec	Zone reconnaissance lane

TRAINING SITE MANEUVER AREAS AND GUNNERY CAPABILITIES

In this appendix we discuss in greater detail the maneuver area capabilities for the installations selected as possible brigade training sites in Chapter Three. We also discuss why two installations that were not selected did not meet maneuver area requirements.

MANEUVER AREA REQUIREMENTS

There are two primary sets of maneuver area requirements, company lanes and brigade and battalion operations. These requirements are shown in Table B.1. The requirements for company lanes are somewhat smaller than those outlined in Department of the Army Training Circular (TC) 25-1 but are seen as adequate for the more structured nature of a lane training event, which is designed to be a limited segment of an operation rather than an entire operation. Battalion and brigade maneuver area requirements are for a complete operation and are the same as outlined in TC 25-1.

Using these requirements as initial guides, we looked at tactical scale maps to draw out areas or mobility corridors that would permit conduct of the training event. The area selected had to reasonably allow the tasks in the lane or FTX to be performed (and thus reach the training objectives) by allowing mounted maneuver with a reasonable level of tactical deployment and providing adequate fields of fire. For example, the area chosen for the tactical move/hasty attack had to have an area that allowed movement down the long axis adequate for a company team tactical formation and an OPFOR platoon hasty defensive position. Even if the area was smaller than the initial guidelines listed above but allowed a successful training event, we considered it satisfactory.

INSTALLATIONS

In all installations we examined, we found that the brigade and battalion maneuver areas were the more difficult criteria to meet.

Table B.1

Maneuver Area Requirements

Event	Area (km)[a]
Company lane/STX	
Hasty defense/reposition	3×7
Deliberate defense	3×5
Counter reconnaissance	6×3
Tactical movement/hasty attack	3×6
Breach	2×4
Assault	2×4
Advance guard/support by fire	3×6
Trench	2×4
Cavalry troop	4×12
Battalion task force FTX	
Defense in sector	6×23
Deliberate attack	4×17
Movement to contact	8×31
Brigade FTX	
Deliberate attack	8×31

[a]Includes area for both BLUEFOR and OPFOR.

Fort Bliss has a massive maneuver area of over 500 square acres, most of which is usable maneuver area. The only major maneuver limitation is the fact that the maneuver areas are 20 miles distant from the permanent areas, and, similar to Piñon Canyon at Fort Carson, this would require greater transportation support and establishment of a corps support unit organization.

Fort Irwin has the battalion and brigade maneuver areas shown on the map in Figure B.1. The CALFEX can be conducted concurrently with the use of both maneuver boxes. There is also adequate maneuver area to conduct nine concurrent company team lanes.

Fort Hood has two maneuver boxes, shown on Figure B.2. Note that the long axes of these boxes are less than the dimensions shown on Table B.1 for the battalion FTX of movement to contact and defense. Still, the terrain is sufficient for the FTX both to be executed and to meet training objectives. Both maneuver areas could be used during the brigade operations to meet the requirement for that brigade's deliberate attack FTX. There is sufficient maneuver area to meet the company team requirements. As we discussed in Chapter Three, there are other constraints. These maneuver areas are in close proximity to built-up areas and civilian roads. Effective training—in our view and in that of several army trainers with whom we discussed this issue—is possible, but this site has important constraints compared to the other four installations selected.

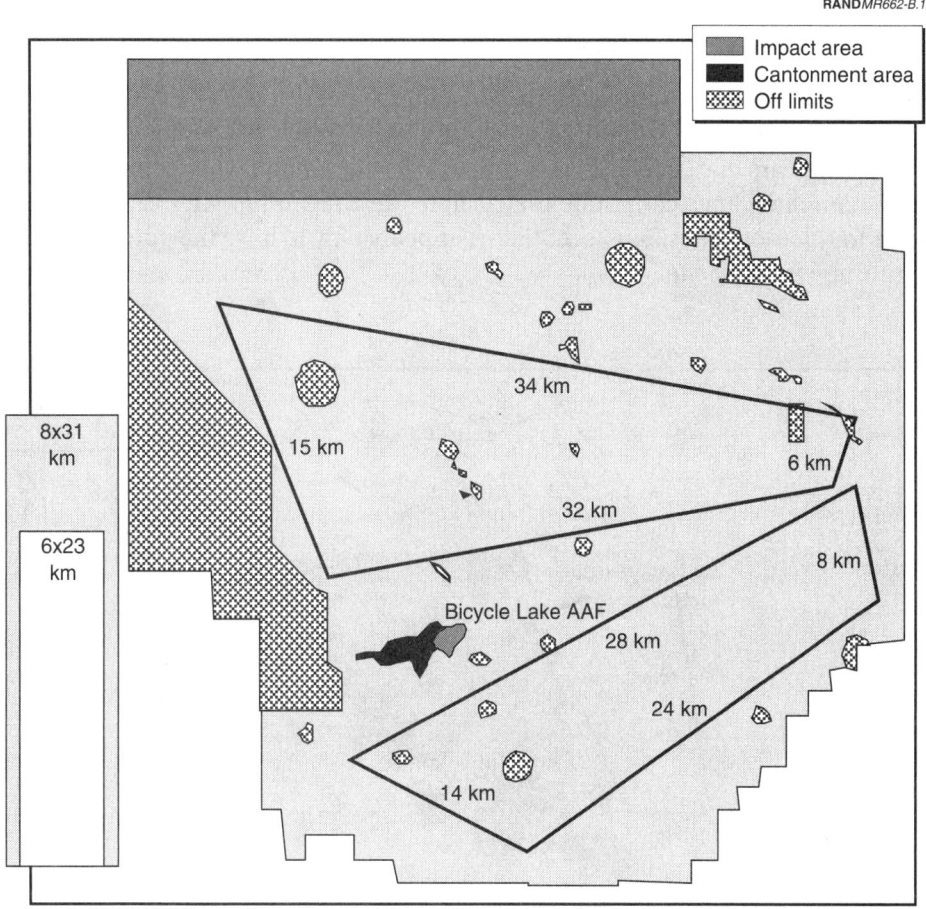

RAND*MR662-B.1*

Impact area
Cantonment area
Off limits

8x31 km

6x23 km

15 km

34 km

6 km

32 km

8 km

Bicycle Lake AAF

28 km

24 km

14 km

Figure B.1—Maneuver Areas at Fort Irwin

Fort Carson/Piñon Canyon. The maneuver areas at Piñon Canyon are shown on Figure B.3. Again, the long axis of one of the boxes is shorter than the guidance found in TC 25-1, but somewhat larger than Fort Hood's. A review of the maneuver areas shows that FTX can meet the training objectives. Again, there is ample area for company lanes.

Fort Riley's possible battalion maneuver area is shown in Figure B.4. Note that there is only one battalion maneuver box and that its long axis would not meet the TC 25-1 guidance for movement to contact or defend in sector operations. Its maneuver area also would not be adequate for a brigade deliberate attack. We were not able to lay out the nine company team/cavalry lanes that could run simultaneously that meet the area criteria shown in Table B.1, although it is likely that with a more detailed exercise design the terrain could be utilized to accomplish company-level training. By our criteria, Fort Riley is probably adequate as a company training site but not for brigade and battalion FTX.

Fort Stewart has only one maneuver area that appears usable for mounted maneuver training; with its limited fields of fire and room for deployment, however, it appears marginal even for company-level training. The requirement for nine company lanes cannot be met. There are a few other sites that could be used for platoon lanes, but Fort Stewart appears only to have the potential to be a gunnery training site.

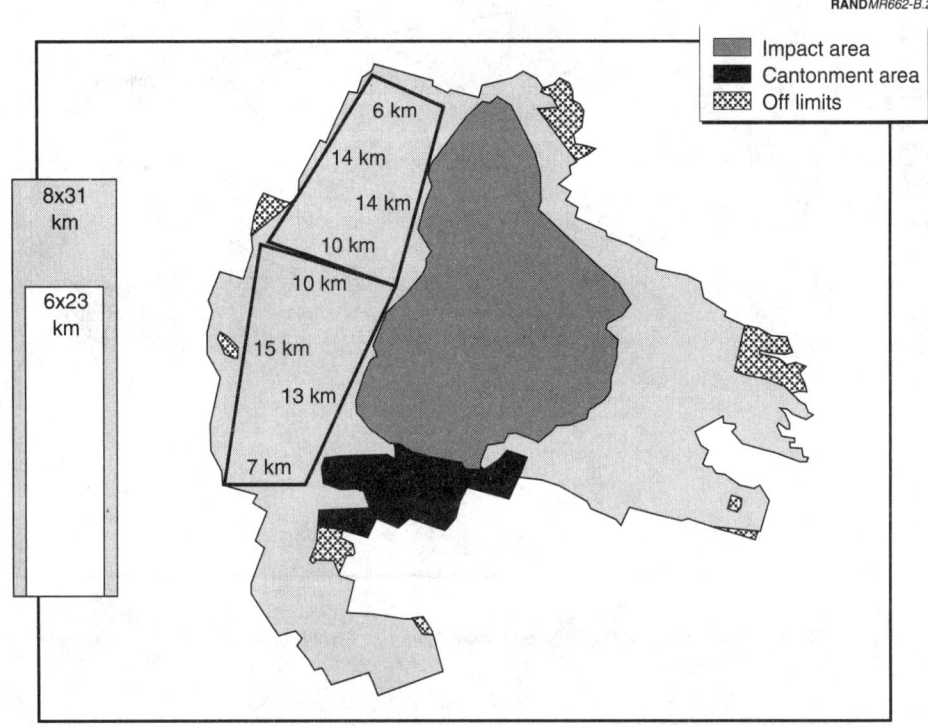

Figure B.2—Maneuver Areas at Fort Hood

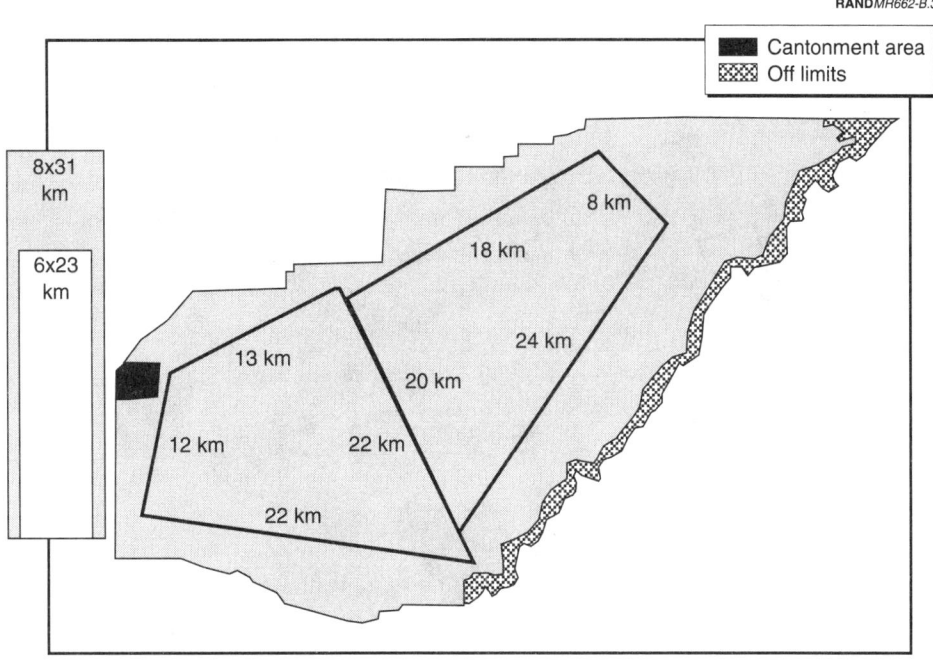

Figure B.3—Maneuver Areas at Piñon Canyon

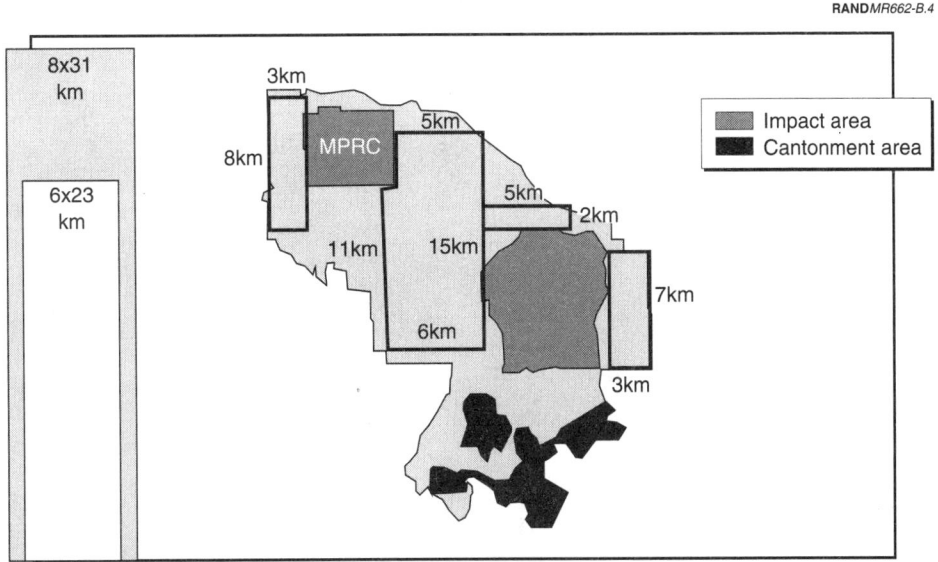

Figure B.4—Maneuver Areas at Fort Riley

RAND*MR662-B.5*

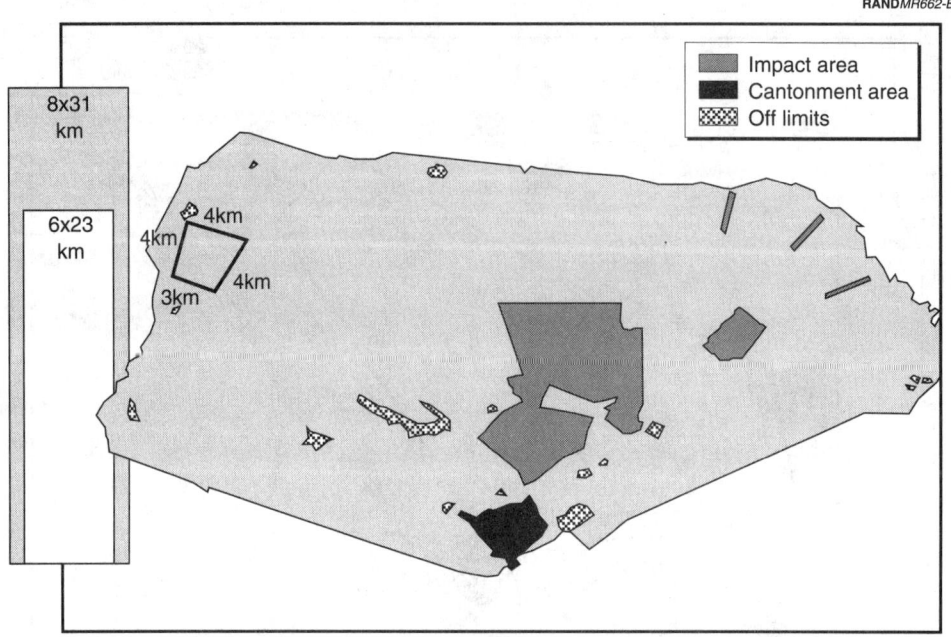

Figure B.5—Maneuver Areas at Fort Stewart

TRAINER, TRAINING MANAGER, AND TRAINING SUPPORT REQUIREMENTS

This appendix details how the trainers, training managers, and training support personnel are organized to support the training model at a single site. We describe the function of each division, section, and team of the training organization and the number, grade, and MOS or branch of each person assigned. We also identify the positions required to be manned by experienced subject matter experts (SMEs). (The positions requiring subject expertise are marked with an asterisk in the tables.)

As we discussed in Chapter Three, the basis for this organization is NTC Operations Group, which we supplemented to perform additional training activities. These activities were associated with lanes and gunnery training events and types of training the NTC is not at present staffed to provide but which is appropriate for National Guard units during their transition to active status, such as individual and section SIDPERS and TAMMS training.

In the tables below we divide the training teams into two categories of trainers and support. The first category is labeled "permanent trainers," that is, trainers and training control personnel who are associated with an element of the brigade and continuously accompany and work with it from Day 13 through the end of training. The second category is designated "additional training personnel," that is, those trainers, training managers, and training support personnel needed to supplement the permanent training personnel for selected events such as gunnery ranges.

In the tables describing the training organizations, we show the number of personnel needed for each unit, range, or lane; number of units, ranges, or lanes at each site; number of personnel needed at each site; and the number of personnel needed as it affects the total requirement. In some cases, the number of personnel for the total site requirement is less than that reflected for each unit, range, and lane because we shift available personnel with the appropriate qualifications to different events to reduce the total requirement as far as possible.

The basic training organization is shown in Figure C.1, along with the personnel requirements in parentheses.

TRAINING HEADQUARTERS

This organization was modeled on organizations with similar functions in the NTC Operations Group and consists of a Command Section, Plans and Operations Division, and Support Division.

- **Command Section: Days 13–102.** This section controls the training operation and coordinates training requirements with supporting installation, supporting National Guard, and OPFOR organizations. It commands the AC training organization and provides internal administrative support.

- **Plans and Operations Division: Days 13–102.** Develops training scenarios and schedules, coordinates maneuver areas and gunnery range usage and coordinates day to day training activities. This section also operates the Training Net Control Station (NCS) on a two-shift, 24-hour basis.

- **Support Division: Days 13–102.** This section provides internal sustainment support for assigned trainer personnel.

RAND*MR662-C.1*

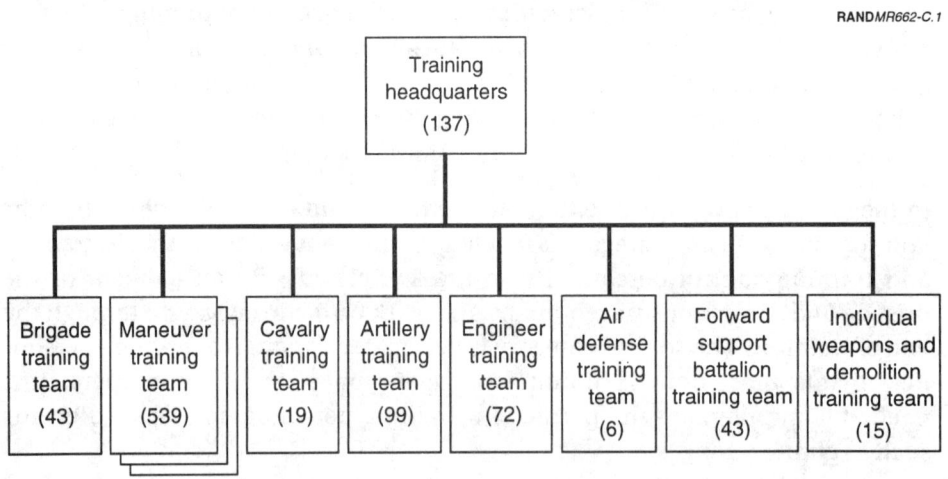

Figure C.1—Training Site Organization

Table C.1

Command Section Requirements

Position	SME	Rank/ Grade	MOS	# per Unit	# of Units per Site	Required # per Site
Command Group						
Commander	*	COL	AR	1	1	1
XO	*	LTC	AR	1	1	1
CSM	*	E9	00Z	1	1	1
RTO /Driver	*	E5	11M/19K	3	1	3
Total						6
Personnel Team						
Chief S-1	*	MAJ	AR	1	1	1
PS NCO	*	E7	75Z	1	1	1
Finance NCO	*	E6	73C	1	1	1
Pers Act NCO	*	E6	75B	1	1	1
Admin NCO	*	E5	71L	1	1	1
Pers Admin Clerk	*	E5	75B	1	1	1
Total						6
Command Section Total						12

Table C.2

Plans and Operations Division Requirements

Position	SME	Rank/ Grade	MOS	# per Unit	# of Units per Site	# Required per Site
Intell Section						
Chief S-2	*	MAJ	MI	1	1	1
Intel NCO	*	E7	96B	1	1	1
Total						2
Operations						
Chief S-3	*	LTC	OPS	1	1	1
Ops Officer	*	MAJ	IN	1	1	1
Ops Officer	*	MAJ	AR	1	1	1
Ops Officer	*	MAJ	FA	1	1	1
Ops Officer	*	CPT	IN	1	1	1
Ops Officer	*	CPT	FA	1	1	1
Ops Officer	*	CPT	EN	1	1	1
Ops Officer	*	CPT	CM	1	1	1
Safety Officer	*	WO	153A	1	1	1
Ops SGM	*	E9	00Z	1	1	1
Ops NCO	*	E9	00Z	1	1	1
Plans NCO	*	E7	11M	1	1	1
Plans NCO	*	E7	12B	1	1	1
Plans NCO	*	E7	13B	1	1	1
Plans NCO	*	E7	14S	1	1	1
Plans NCO	*	E7	19K	1	1	1
NBC NCO	*	E7	54B	1	1	1
Plans NCO	*	E7	96B	1	1	1
Admin Spec	*	E5	71C	1	1	1
Total						19
Plans and Operations Total						21

Table C.3

Support Division Requirements

	SME	Rank/ Grade	MOS	# per Unit	# of Units per Site	Required # per Site
Logistics						
Chief S-4	*	MAJ	LOG	1	1	1
Asst S-4	*	CPT	OD	1	1	1
Prop Bk Officer	*	WO	920A	1	1	1
Supp Tech	*	WO	920B	1	1	1
NCOIC	*	E8	92Z	1	1	1
Supp NCO	*	E7	92Y	1	1	1
Asst Supp NCO	*	E6	92Y	1	1	1
Admin Supp	*	E5	71L	1	1	1
Total						8
Services						
Mess Team		E6	94B	1	1	1
		E5	94B	3	1	3
		E4	94B	13	1	13
Supply Team		WO	Supply	1	1	1
		E7	76Y	1	1	1
		E5	76Y	2	1	2
		E4	76Y	3	1	3
		E4	77F	3	1	3
		E4	55B	2	1	2
Maintenance Team		WO	Maint	1	1	1
		E7	63B	1	1	1
		E6	31V	1	1	1
		E6	63B	5	1	5
		E5	63B	5	1	5
		E4	31V	3	1	3
		E4	63B	20	1	20
Medical Team		WO	Med	1	1	1
		E7	91B	1	1	1
		E6	94B	2	1	2
		E5	94B	5	1	5
		E4	91B	10	1	10
Trans Team		E7	88M	1	1	1
		E6	88M	2	1	2
		E5	88M	2	1	2
		E4	88M	7	1	7
Total						96
Support Division Total						104
Training HQS Total						137

BRIGADE TRAINING TEAM

This organization was modeled on the NTC Operations Group's Brigade Trainers or "Broncos." It is somewhat larger in that we added trainers to provide greater coverage of supply and maintenance functions. It consists of the following groups of permanent training personnel.

- **Brigade command and control trainers: Days 13–102.** These trainers align with the brigade commander and his primary and key special staff sections. Their primary responsibility is to provide training feedback and advice to their counterparts, starting with individual and section-level training during the preliminary command and control training phase and continuing through completion of final preparations for deployment. They also assist with training control and enforcement of rules of engagement. Fire support training personnel are not included here but are listed under the FA brigade.

- **Brigade HHC and special platoon trainers: Days 13–102.** These trainers align with the HHC commander, first sergeant and executive officer, the mechanized infantry (MI) company commander, and leaders of the special platoons, and they are responsible for providing training feedback and advice. These trainers remain with the organization for the duration of training starting on Day 13.

- **Brigade net control station (NCS) section: Days 13–102.** This section portrays the divisional (or corps) nets, monitors the activities on the brigade and divisional nets, tracks ongoing operations for the trainers, and provides data to support AARs and other training requirements. It supports preliminary C2 and FTX training. The NCS operates 24 hours a day on a two-shift schedule.

Table C.4

Brigade C2 Trainer Requirements

Position	SME	Rank/Grade	MOS	# per Unit	# of Units per Site	Required # per Site
Senior Tnr	*	LTC/COL	IN/AR	1	1	1
Asst. BDE	*	LTC	AR/IN	1	1	1
Sr. NCO	*	SGM	11Z	1	1	1
S1	*	MAJ	AG/12A	1	1	1
SIDPERS	*	E7	75Z	1	1	1
S2	*	MAJ	MI	1	1	1
Asst. S2	*	CPT	MI	1	1	1
S3	*	MAJ	AR/IN	1	1	1
Asst. S3	*	CPT	IN/AR	1	1	1
Ops	*	MSG	11M	1	1	1
S4	*	MAJ	AR/IN	1	1	1
Asst. S4	*	E7	76Y	1	1	1
Chemical	*	CPT	CM	1	1	1
Signal	*	MAJ	SIG	1	1	1
Asst. Signal	*	E7	31W	1	1	1
ADA	*	MAJ	ADA	1	1	1
Engineer	*	MAJ	EN	1	1	1
Total						17

NOTE: During C2 simulations, additional training support personnel are needed to operate the simulations. These would be provided by the installation or RC personnel from the BCST of the USAR's Divisions Exercise but are not included in our requirements.

Table C.5

Brigade HHC and Special Platoon Trainer Requirements

Position	SME	Rank/Grade	MOS	# per Unit	# of Units per Site	Required # per Site
CO	*	MAJ/CPT	AR/IN/12A	1	1	1
XO/1SG	*	E8	11M	1	1	1
Smoke	*	E7	54B	1	1	1
Decon	*	E7	54B	1	1	1
Chem. Rcn	*	E7	54B	1	1	1
MP	*	E8/E7	95Z/95B	1	1	1
Total						6

Table C.6

Brigade NCS Requirements

Position	SME	Rank/ Grade	MOS	# per Unit	# of Units per Site	Required # per Site
OIC	*	MAJ	IN/AR	2	1	2
Asst. OIC	*	CPT	AR/IN	2	1	2
NCOIC	*	E8	11M/19K	2	1	2
Ops Intel (G2)	*	E8	96B	2	1	2
Command (G3)	*	E8	19K/11M	2	1	2
Admin/Log	*	E8	76Y/75Z	2	1	2
FA (DivArty)	*	CPT	FA	2	1	2
Drivers		E4	11M/19K	4	1	4
Total						20
Brigade Training Team Total						43

MANEUVER BATTALION TASK FORCE TRAINING TEAMS

These organizations are modeled on the NTC Operations Group's Maneuver Battalion Training teams, or "Scorpions, Cobras, Dragons, and Tarantulas." It is somewhat larger in that we added some trainers for brigade sustainment functions (e.g., SIDPERS and company executive officer/first sergeant trainers). Figure C.2 shows the elements of a typical maneuver battalion training team. The number of company training teams will vary depending on the makeup of the unit. We first list the permanent training personnel and then the additional maneuver battalion gunnery and lane trainers, managers, and personel.

Permanent Training Personnel

- **Battalion task force command and control trainers.** These trainers align with the battalion task force commanders and their primary and key special staff sections. Their primary responsibility is to provide training feedback and advice to their counterparts, starting with individual and section-level training during preliminary command and control training phase and continuing through completion of final preparations for deployment. They also assist with training control and enforcement of rules of engagement. Fire support training personnel are not included in this chart but are listed under the FA brigade.

- **Battalion task force HHC trainers.** These trainers align with the HHC commander, first sergeant and executive officer, and support platoon. They provide training control, feedback, and advice.

- **Maneuver company/company team trainers.** These trainers align with the company commander, first sergeant, and executive officer, and each maneuver platoon. They provide training feedback and advice. They provide a permanent set of trainers for each company and platoon. They also assist with control training and enforcement of rules of engagement.

- **Maneuver battalion scout platoon and mortar platoon trainers.** These trainers align with the platoon leader and sections of these platoons. They provide training feedback and advice. They also assist with control training and enforcement of rules of engagement. The mortar platoon trainers also cover the mortar sections in the cavalry troop.

- **Battalion task force net control section (NCS).** The NCS section monitors the activities on the brigade and battalion nets, tracks ongoing operations for the trainers, and provides data to support AARs and other training requirements. All NCS operate on a two-shift basis. FSS trainers are shown under the FA battalion.

RAND*MR662-C.2*

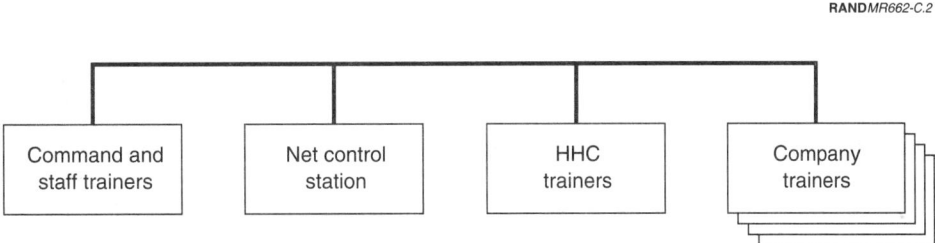

Figure C.2—Typical Battalion Training Team

Table C.7

Tank Battalion Command and Staff Requirements

Position	S M E	Rank/ Grade	MOS	# per Unit	# of Units per Site	Required # per Site
Senior Trainer	*	LTC	AR	1	1	1
XO/CSS	*	MAJ	AR	1	1	1
S1	*	CPT	AR	1	1	1
SIDPERS	*	E8/E7	75Z	1	1	1
S2	*	CPT	MI	1	1	1
S3	*	MAJ	AR	1	1	1
Asst. S3	*	CPT	AR	1	1	1
Combat Engr	*	CPT	EN	1	1	1
S4	*	CPT	AR	1	1	1
Asst. S4	*	E7/E6	76Y/92Y	1	1	1
ADA	*	CPT	ADA	1	1	1
Maint.	*	CPT	AR	1	1	1
TAMMS	*	WO/E7	915D/63E	1	1	1
Medical	*	CPT	MSC	1	1	1
Asst. Medical	*	E7	91B	1	1	1
NBC	*	CPT	CM	1	1	1
Asst. NBC	*	E7	94B/54B	1	1	1
Senior NCO	*	E8	19K/19D	1	1	1
Staff Total						18

NOTE: During C2 simulations, additional training support personnel are needed to operate the simulations. These would be provided by the installation or from RC personnel from the BCST of the USAR's Divisions Exercise, but are not included in our requirements.

Table C.8

Mechanized Infantry Battalion Command and Staff Requirements

Position	SME	Rank/ Grade	MOS	# per Unit	# of Units per Site	Required # per Site
Senior Battalion Trainer	*	LTC	IN/AR	1	2	2
XO/CSS	*	MAJ	IN	1	2	2
S1	*	CPT	IN	1	2	2
SIDPERS	*	E7	75Z	1	2	2
S2	*	CPT	MI	1	2	2
S3	*	LTC/MAJ	IN	1	2	2
Asst. S3	*	CPT	IN/AR	1	2	2
Combat Engr	*	CPT	EN	1	2	2
S4	*	CPT	IN	1	2	2
Asst S4	*	E7/E6	76Y/92Y	1	2	2
ADA	*	CPT	ADA	1	2	2
Maint.	*	CPT	IN	1	2	2
TAMMS	*	WO/E7	915D/63T	1	2	2
Medical	*	CPT	MSC	1	2	2
Asst. Medical	*	E7	91B	1	2	2
NBC	*	CPT	CM	1	2	2
Asst, NBC	*.	E7	94B/54B	1	2	2
Senior NCO	*	E9/E8	11Z/11M	1	2	2
Staff Total						36

NOTE: During C2 simulations, additional training support personnel are needed to operate the simulations. These would be provided by the installation or RC personnel from the BCST of the USAR Divisions Exercise, but are not included in our requirements.

Table C.9

Tank Battalion HHC Trainer Requirements

Position	SME	Rank/ Grade	MOS	# per Unit	# of Units per Site	Required # per Site
CO Trainer	*	CPT	AR	1	1	1
XO/1SG	*	CPT/E8	AR/19K	1	1	1
Support Plt	*	E7	88M/19K	1	1	1
HHC Total						3

Table C.10

Mechanized Infantry HHC Trainer Requirements

Position	S M E	Rank/ Grade	MOS	# per Unit	# of Units per Site	Required # per Site
CO Trainer	*	CPT	IN	1	2	2
XO/1SG	*	E8	11M	1	2	2
Support Plt	*	E8/E7	88M/11M	1	2	2
HHC Total						6

Table C.11

Tank Company Trainer Requirements

Position	S M E	Rank/ Grade	MOS	# per Unit	# of Units per Site	Required # per Site
CO	*	CPT	AR	1	4	4
XO/1SG	*	CPT/E8	AR/19K	1	4	4
PLT	*	E7	19K	3	4	12
Total						20

Table C.12

Mechanized Infantry Company Trainer Requirements

Position	S M E	Rank/ Grade	MOS	# per Unit	# of Units per Site	Required # per Site
CO	*	CPT	IN	1	8	8
XO/1SG	*	CPT/E8	IN/11M	1	8	8
PLT	*	E7	11M	3	8	24
Asst. PLT	*	E7/E6	11M	3	8	24
Total						64

Table C.13

Scout Platoon Trainer Requirements

Scout Platoon Trainers	SME	Rank/ Grade	MOS	# per Unit	# of Units per Site	Required # per Site
PLT	*	CPT	AR/IN	1	3	3
Asst. PLT	*	E7	19D	1	3	3
Section	*	E7/E6	19D	1	3	3
Total						9

NOTE: We assume that the scout platoon will have 10 HMMWVs and 4 sections.

Table C.14

Mortar Platoon Trainer Requirements

Mortar Platoon Trainers	SME	Rank/ Grade	MOS	# per Unit Dry Fire	# per Unit Live Fire	# of Units per Site	Required # per Site
PLT	*	CPT/E8/E7	IN/11C	1	1	3	3
SEC	*	E7/E6	11C	1	1	3	3
FDC	*	E7/E6	11C	1	1	3	3
CAV TRP	*	E7/E6	11C	1	1	1	1
Total							10

Table C.15

Tank Battalion NCS Requirements

Range Management and Support	SME	Rank/ Grade	MOS	# per Unit	# of Units per Site	# per Site	# Required to Source
OIC	*	LTC/MAJ	AR	2	1	2	2
NCOIC	*	E8	19K	2	1	2	Covered by
Ops Intel	*	E7	19K	2	1	2	Additional
Command	*	E7	19K	2	1	2	Gunnery
OC/OC Cmd.	*	E7	19K	2	1	2	O/C-Ts
Drivers		E4		4	1	4	
Admin/Log	*	E7/E6	75Z/76Y/92Y	2	1	2	2
FA LNO	*	E7/E6	13F	2	1	2	2
NCS Total							6

Table C.16

Mechanized Infantry Battalion NCS Requirements

Range Management and Support	S M E	Rank/ Grade	MOS	# per Unit	# of Units per Site	# per Site	# Required to Source
OIC	*	LTC/MAJ	IN/AR	2	2	4	4
NCOIC	*	E8	11M/51Z	2	2	4	Covered by
Ops Intel	*	E7	11M	2	2	4	Additional
Command	*	E7	11M	2	2	4	Gunnery
OC/OC Cmd.	*	E7	11M	2	2	4	O/C-Ts
Drivers		E4		4	2	8	
Admin/Log	*	E7/E6	76Y/92Y	2	2	4	4
FA LNO (Mortar)	*	E7/E6	13F	2	2	4	4
NCS Total							12

NOTE: During the gunnery period for each of the battalions, NCS tank or Bradley NCO from each battalion support gunnery requirements. Available personnel are shifted to support the NCS preliminary C2 training requirements. Our analysis indicates that this is feasible during these periods.

Additional Maneuver Battalion Gunnery and Lane Trainers, Managers, and Support Personnel

The personnel listed below are only required for the training events and time frames shown.

- **Tank and Bradley gunnery ranges support.** These personnel support both tank and Bradley gunnery training.

- **Preliminary gunnery trainers.** These trainers assist in conducting PMCS and COFT training. COFT training is conducted as a concurrent/remedial training event through the end of Table VIII.

- **Squad live-fire trainers.** The additional trainers augment the permanent platoon O/C-Ts to allow two O/C-Ts per squad during the actual live-fire runs. This coverage is designed to provide a normal complement of trainers for crawl and walk phases and is augmented to cover run phases. This strategy is sufficient to allow one company to complete the LFX in three days.

- **Scout platoon gunnery trainers, management and support.** These requirements are based on a HMMWV scout platoon conducting gunnery in accordance with FM 17-12-8 (Draft). Tables I and II are daylight only; III through X are day and night and require two shifts.

- **Tank gunnery lanes management and support.**

- **Tank platoon lane management and support.**

- **Infantry platoon lanes management and support.**

- **Scout platoon lanes management and support.**

- **Mortar platoon lane trainers, management and support.** Mortar platoon lanes include live-fire exercises, and these personnel cover the mortar sections in the cavalry troop as well as the maneuver battalions. FIST support trainers are also required, but they are listed under the FA battalion.

- **Company team lane management and support.** The personnel below support both mech infantry and tank company teams lanes. Permanent company team and FA FIST perform trainer functions on company lanes.

- **Battalion and brigade operations support personnel.**

Table C.17

Tank and Bradley Gunnery Range Support Requirements (Days 15–49)

Personnel	S M E	Rank/ Grade	MOS	# per Range	# of Ranges per Site	Required # per Site
Spotter	*	E6/E5	11M/19K	2	7	14
Tower/Target Operator	*	E7/E6	11M/19K	2	7	14
Range Support		E6	11M/19K	2	7	14
Target NCO		E6/E5	11M/19K	2	7	14
Target detail		E4		4	5	20
		E5		2	5	10
Guards		E4		4	5	20
Driver/RTO		E5	19K	2	1	2
Total						108

NOTE: Range personnel work on a two-shift basis.

Table C.18

Tank Gunnery Range Trainer, Management, and Support Requirements (Days 15–32)

Range Management and Support	SME	Rank/ Grade	MOS	# per Range	# of Ranges per Site	# per Site	# Required to Source
OIC	*	MAJ/CPT	AR	2	1	2	Covered by CO TM Lane Mgt
NCOIC	*	CPT/E8	19K	2	4	8	Covered by Permanent CO O/C-Ts
Master Gunner	*	E8/E7	19K	1	4	4	
Senior TCE	*	E8/E7	19K	2	4	8	
TCE	*	E7/E6	19K	8	4	32	32
Range Safety	*	E8/E7	19K/19D	2	4	8	8
Script Reader	*	E7/E6/E5	19K	2	4	8	8
Total							48

Table C.19

Bradley Gunnery Range Trainer, Management, and Support Requirements (Days 24–49)

Personnel	SME	Rank/ Grade	MOS	# per Range	# of Ranges per Site	# per Site	# Required to Source
OIC	*	MAJ/CPT	IN	2	1	2	2
NCOIC	*	CPT/E8	IN/11M/11H	2	4	8	Covered by Permanent CO O/C-Ts
Master Gunner	*	E8/E7	11M	1	4	4	
Senior BCE	*	E8/E7	11M	2	4	8	
BCE	*	E7/E6	11M	8	4	32	32
Range Safety	*	E8/E7	11M/19D	2	4	8	8
Script Reader	*	E6/E5	11M/11H/19D	2	4	8	8
Total							50

NOTE: Bradley Table XII requires four additional infantry squad trainers.

Table C.20

Bradley Gunnery Support Requirements (Days 13–30)

Position	SME	Rank/ Grade	MOS	# per Range	# of Ranges per Site	# per Site	# Required to Source
M1 PMCS (Days 13–21)	*	E7/E6	19K	7	1	7	Covered by Permanent CO O/C-Ts
M1 COFT (Days 13–30)	*	E7/E6	19K	10	1	10	
Total							10

Table C.21

Mechanized Infantry Preliminary Gunnery Trainer Requirements
(Days 13–47)

Position	SME	Rank/Grade	MOS	# per Range	# of Ranges per Site	# per Site	# Required to Source
M2/M3 PMCS (Days 13–19)	*	E7/E6	11M	7	1	7	Permanent CO O/C-Ts
M2/M3 COFT (Days 13–47)	*	E7/E6	11M	10	1	10	10
Total							10

Table C.22

Squad Live-Fire Trainer Requirements (Days 20–47)

Position	SME	Rank/Grade	MOS	# per Range	# of Ranges per Site	Required # per Site
OIC	*	CPT	IN	2	2	4
NCOIC/Safety	*	CPT/E8/E7	IN/11M	2	2	4
Squad O/C-Ts	*	E7/E6	11M	1	2	2
Total						10

NOTE: Support requirements are not available in TC 7-9. The support requirements in this table are based on similar tank and Bradley gunnery events.

Table C.23

Scout Platoon Gunnery Trainer Requirements (Days 41–56)

Position	SME	Rank/ Grade	MOS	# per Range	# of Ranges per Site	# per Site	# Required to Source
Range/Lane Support Gunnery Ranges							
Spotter	*	E4	19D/19K/11M	2	1	2	
Tower/Target Opr.	*	E6	19D/19K/11M	2	1	2	Covered by
Range Safety	*	E7	19D	2	1	2	Table V
Range Support		E6	19D/19K/11M	2	1	2	AR/IN
Driver/RTO		E5	19D	1	1	1	Table
Target NCO		E6	19D/19K/11M	1	1	1	Gunnery
Target Detail		E4		4	1	4	Support
Sgt of the Guard		E5		2	1	2	Personnel
Guards		E4		4	1	4	
Additional Scout Gunnery O/C-Ts (Days 41–56)							
Gunnery	*	E7/E6	19D	3	1	3	3
Total							3

NOTES: All scouts fire the same table on the same day; tables are sequential.

Table C.24

Squad Live-Fire Range Support Requirements (Days 26–49)

Position	SME	Rank/ Grade	MOS	# per Range	# of Ranges per Site	Required # per site
Driver/RTO		E5	11M	2	2	4
Spotter	*	E6/E5/E4	11M	2	2	4
Tower/Target Operator	*	E7/E6	11M	2	2	4
Range Support NCO	*	E7/E6	11H/11M	1	2	2
Target Detail		E7/E5	12B	2	2	4
Total						18

Table C.25

Tank Platoon Lane Management and Support Requirements (Days 33–47)

Position	S M E	Rank/ Grade	MOS	# per Lane	# of Ranges per Lane	# per Site	# Required to Source
Lane Management							
OIC	*	CPT	AR	1	4	4	Covered by
NCOIC	*	E8/E7	19K	1	4	4	CO TM Lane Mgt
Lane Support							
Fire Markers		E5		2	4	8	8
Drivers/RTO		E5	19K	1	4	4	4
General		E4		2	4	8	8
Total							20

Table C.26

Infantry Platoon Lane Management and Support Requirements (Days 22–49)

Position	S M E	Rank/ Grade	MOS	# per Lane	# of Lanes per Site	# per Site	# Required to Source
Lane Management							
OIC	*	CPT	IN	1	4	4	Covered by Permanent
NCOIC	*	E8/E7	11M/11H	1	4	4	CO O/C-Ts
Lane Support							
Fire Markers		E5		2	4	8	8
Drivers/RTO		E5	11M	1	4	4	4
General		E4		2	4	8	8
Total							20

Table C.27

Scout Platoon Lane Management and Support Requirements (Days 20–34)

Position	S M E	Rank/ Grade	MOS	# per Lane	# of Lanes per Site	# per Site	# Required to Source
NCOIC	*	E8/E7	19D	1	1	1	Covered by
Fire Markers		E5		2	1	2	Add. Gunnery
Drivers/RTO		E5	19D	1	1	1	O/C-Ts
General		E4	11B	2	1	2	2
Total							2
Scout Platoon Total							14

Table C.28

Mortar Platoon Lane Management and Support Requirements (Days 13–46)

Position	SME	Rank/ Grade	MOS	# per Lane Dry Fire	# per Lane Live Fire	# of Lanes per Site	# per Site	# Required to Source
Mortar PLT Lane Management								
OIC	*	CPT	IN	1	1	1	1	Covered by Permanent O/C-Ts
NCOIC	*	E7	11C	1	1	1	1	
Range Safety	*	CPT/E8/E7	IN/11C		1	1	1	
Mortar PLT Lane O/C-Ts								
PLT	*	CPT/E8/E7	IN/11C	1	1	1	1	Covered by Permanent O/C-Ts
Asst PLT	*	E7/E6	11C	1	2	1	2	
SEC	*	E7/E6	11C	2	2	1	2	
FDC	*	E7/E6	11C	2	2	1	2	
Mortar PLT Lane Support								
Driver		E6/E5	11C	1	1	1	1	1
Range Support		E6	11C		1	1	1	1
Sgt of the Guard		E5			1	1	1	1
Guards		E4			2	1	2	2
Total								5
Mortar Platoon Total								15

Table C.29

Company Team Management and Support Requirements (Days 52–67)

Position	SME	Rank/ Grade	MOS	# per Lane	# of Lanes per Site	# per Site	# Required to Source
Company Team Management							
OIC	*	CPT	IN/AR	1	9	9	9
NCOIC	*	E7	11M/19K	1	9	9	9
NCS	*	E7/E6	11M/19K	1	9	9	9
Total							27
Company Team Support							
Smoke Gen. (1)	*	E5	54B	1	4	4	4
	*	E4	54B	1	4	4	4
Fire Markers		E6/E5		2	9	18	Covered by Gunnery Range Support
Drivers/RTO		E5	11M/19K	1	9	9	
General		E4		2	9	18	
Total							8
Company Team Management Total							35

Table C.30

Battalion Task Force and Brigade Operations Support Requirements (Days 69–92)

Position	SME	Rank/ Grade	MOS	# per Event	# of Events per Site	# per Site	# Required to Source
Smoke Gen. (4)	*	E5	54B	6	2	12	8
	*	E4	54B	6	2	12	8
Fire Markers		E5		12	2	24	Covered by
Drivers/RTOs		E5/E4	11M/19K/19D	6	2	12	CO TM Lane
General		E4		6	2	12	Support
Total							16

CALFEX MANAGEMENT AND SUPPORT

There are two company-level CALFEX, movement to contact and defend in sector. The exercises are conducted with an appropriate portion of the CS slice in support of the company team, engineers, FA, mortars, and scouts. The battalion task force controls the company team. The table below describes only CALFEX management and support personnel requirements. Trainer requirements are covered by the permanent trainers for each element participating in the exercise. The requirement for these personnel is less than required to support tank and Bradley gunnery, and because CALFEX are conducted after gunnery, gunnery management and support personnel can be shifted to cover CALFEX requirements.

Table C.31

Tank Platoon Lane Management and Support Requirements (Days 69–88)

Position	S M E	Rank/ Grade	MOS	# per Range	# of Ranges per Site	# per Site	# Required to Source
Overall Control							
CALFEX OIC	*	MAJ	AR or IN	1	1	1	
Driver/RTO		E4	19K or 11M	1	1	1	
Range Management							
Range OIC	*	CPT	AR or IN	1	2	2	Covered by
NCOIC	*	E8	11M or 19K	1	2	2	AR & IN
Net Control Station	*	E7	11M or 19K	2	2	4	Gunnery
Range Support							
Spotter	*	E4	11M or 19K	1	2	2	
Tower Operator	*	E6	11M or 19K	1	2	2	
Range Safety Off.	*	CPT	IN or AR	1	2	2	
Range support		E6	11M or 19K	1	2	2	
Target NCO	*	E6	11M or 19K	1	2	2	
Target detail		E4		2	2	4	
Sgt of the Guard		E5		1	2	2	
Guards		E4		3	2	6	

NOTES: Cav Trp executes one event at a time (DS followed by Route Recon). One half of personnel are required (2 + 16 = 18) to support CAV TRP Live Fire.

FIELD ARTILLERY BATTALION TRAINING TEAM

This organization was modeled on the NTC Operations Group's field artillery battalion trainers or "Werewolves." It has been augmented to provide additional section and individual capability. The coverage below includes requirements to cover FIST and FSS trainers. This team also provides fire marker control and support requirements during brigade and battalion operations.

Permanent Training Personnel

- **FA battalion command and control trainers.** These trainers align with the battalion commander and the battalion's primary and key special staff sections. Their primary responsibility is to provide training feedback and advice to their counterparts, starting with individual and section-level training through completion of final preparations for deployment. They also assist with training control and enforcement of rules of engagement.

- **FA HHB and service battery trainers.** These trainers align with the HHB commander, first sergeant, and executive officer. They provide training feedback and advice.

- **FA batteries.** These trainers align with the battery commanders, first sergeant and executive officer, and key sections and platoons. They provide training feedback and advice. They provide a permanent set of trainers for each battery and platoon. They also assist with control training and enforcement of rules of engagement.

- **Fire support system trainers.** These trainers have the mission of training the fire support personnel organic to the FA battalion but who are assigned to maneuver battalions during tactical operations. These are the Fire Support Sections (assigned to maneuver battalions and the brigade) and FIST (assigned to maneuver companies and the cavalry troop). These trainers are sufficient to perform fire support system training for preliminary command and control training, mortar training, maneuver platoon, and company lane training as well as battalion and brigade operations.

- **FA battalion net control section (NCS).** This section portrays the divisional (or corps) artillery nets, monitors the activities on the brigade and battalion nets, tracks ongoing operations for the trainers, and provides data to support AARs and other training requirements. All NCS operate on a two-shift basis. The FA battalion NCS also has the mission of controlling fire marker support for lanes and battalion task force and brigade operations training events.

Table C.32

FA Battalion Command and Control Trainer Requirements

Position	SME	Rank/ Grade	MOS	# per Unit	# of Units per Site	Required # per Site
C2 O/C-Ts						
Senior Tnr	*	LTC	FA	1	1	1
XO/CSS	*	MAJ	FA	1	1	1
S1	*	CPT	FA	1	1	1
SIDPERS	*	E7	75Z	1	1	1
S2	*	CPT	MI	1	1	1
S3	*	MAJ	FA	1	1	1
S4	*	CPT	FA	1	1	1
Asst. S4	*	E6/E7	76Y/92Y	1	1	1
Maint.	*	CPT	FA	1	1	1
TAMMS	*	WO/E7	915A/63D	1	1	1
Medical	*	CPT	MSC	1	1	1
Asst. Medical	*	E7	91B	1	1	1
Senior FDC	*	CPT	FA	1	1	1
Commo	*	CPT/E7	SIG/31B	1	1	1
Tac Fire	*	E7	13C	1	1	1
Survey	*	E7	82C	1	1	1
Radar	*	WO/E7	131A/13R	1	1	1
NBC	*	CPT/E7	CM/54B	1	1	1
Senior NCO	*	E9/E8	13Z	1	1	1
Total						19

NOTE: During C2 simulations, additional training support personnel are needed to operate the simulations. These would be provided by the installation or from RC personnel from the BCST of the USAR's Divisions Exercise, but are not included in our requirements.

Table C.33

FA HHB and Service Battery Trainer Requirements

Position	SME	Rank/ Grade	MOS	# per Unit	# of Units per Site	Required # per Site
HHB O/C-Ts						
BTY	*	CPT	FA	1	1	1
XO/1SG	*	CPT/E8	FA/13Z	1	1	1
Total						2
SVS BTY O/C-Ts						
BTY	*	CPT	FA	1	1	1
XO/1SG	*	CPT/E8	FA/13B	1	1	1
Field Trains	*	CPT	FA	1	1	1
Combat Trains	*	CPT	FA	1	1	1
Total						4
HHB/Service Battery Total						6

Table C.34

FA Battery Trainer Requirements

Position	SME	Rank/ Grade	MOS	# per Unit	# of Units per Site	Required # per Site
Battery Trainer	*	CPT	FA	1	3	3
XO/1SG	*	CPT/E8	FA/13Z	1	3	3
PLT/Hwtzr	*	E7	13B	2	3	6
FDC	*	E7/E6	13C/13E	2	3	6
Total						18

Table C.35

Fire Support System Trainer Requirements

Position	SME	Rank/ Grade	MOS	# per Unit	# of Units per Site	Required # per Site
FSO						
BDE	*	MAJ/CPT	FA	2	1	2
BNTF	*	CPT	FA	1	3	3
Total						5
FSE						
BNTF	*	E7	13F	1	3	3
FIST/COLT						
CO/TRP	*	E7/E6	13F	1	15	15
Total						18
Fire Support System Total						23

Table C.36

FA Battalion NCS Requirements

Position	SME	Rank/ Grade	MOS	# per Unit	# of Units per Site	Required # per Site
OIC	*	MAJ	FA	2	1	2
NCOIC	*	E8	13Z	2	1	2
BN/BDE Cmd	*	E7	13F	2	1	2
FD nets	*	E7	13F	2	1	2
FC nets	*	E7	13F	2	1	2
FA BN nets	*	E7	13F	2	1	2
FA Admin/Log	*	E7	76Y/92Y	2	1	2
Fire Mkr.Coord.	*	E7	13F	2	1	2
O/C nets	*	E7	13B	2	1	2
Driver		E4	13F	4	1	4
Total						22

Additional FA Battalion Gunnery and Lane Trainers, Training Managers, and Training Support Personnel

When the FA battalion goes through the artillery tables and the battalion EXEVAL additional trainers, training managers and training support personnel are required. The personnel listed below are needed for Days 13 to 50.

Table C.37

FA Gunnery Management and Support Requirements

Position	SME	Rank/ Grade	MOS	# per Unit	# of Units per Site	Required # per Site
Lane Management (Days 13–50)						
OIC	*	MAJ/CPT	FA	1	1	1
NCOIC	*	E7	13B	1	1	1
Total						2
Lane Support (Days 13–50)						
Drivers/RTO		E5	13	2	3	6
General		E4		1	3	3
Total						9
Total Management and Support						·11
Field Artillery Battalion Total						99

ENGINEER BATTALION TRAINING TEAM

This organization was modeled on the NTC Operations Group's Engineer Trainers or "Sidewinders," with some augmentation to provide additional section and individual training capability. It consists of the following groups of permanent training personnel.

Permanent Training Personnel

- **Engineer battalion command and control trainers.** These trainers align with the battalion commander and the battalion's primary and key special staff sections. Their primary responsibility is to provide training feedback and advice to their counterparts, starting with individual and section-level training through completion of final preparations for deployment. They also assist with training control and enforcement of rules of engagement. These trainers remain with the battalion for the duration of training.

- **Engineer battalion HHC trainers.** These trainers align with the HHC commander, first sergeant and executive officer, and support platoon. They provide training feedback and advice.

- **Engineer company.** These trainers align with the company commander, first sergeant and executive officer, and each platoon. They provide training feedback and advice. They provide a permanent set of trainers for each company and platoon. They also assist with control training and enforcement of rules of engagement.

- **Engineer battalion net control section (NCS).** This section monitors the activities on the brigade and divisional nets, tracks ongoing operations for the trainers, and provides data to support AARs and other training requirements. All NCS operate on a two-shift basis.

Table C.38

Engineer Battalion Command and Control Trainer Requirements

Position	S M E	Rank/ Grade	MOS	# per unit	# of units per site	Required # per site
Senior	*	LTC	EN	1	1	1
XO/CSS	*	MAJ	EN	1	1	1
S1	*	CPT	EN	1	1	1
SIDPERS	*	E7	75Z	1	1	1
S2	*	CPT	MI	1	1	1
S3	*	MAJ	EN	1	1	1
Asst. S3	*	CPT	EN	1	1	1
S4	*	CPT	EN	1	1	1
Maint.	*	WO	EN	1	1	1
TAMMS	*	WO/E7	915A/62B/63H	1	1	1
Medical	*	CPT	MSC	1	1	1
Asst. Medical	*	E7	91B	1	1	1
NBC	*	CPT/E7	CM/54B	1	1	1
Senior NCO	*	E9/E8	12B/12Z	1	1	1
Total						14

NOTE: During C2 simulations, additional training support personnel are needed to operate the simulations. These would be provided by the installation or from RC personnel from the BCST of the USAR's Divisions (Exercise), but are not included in our requirements.

Table C.39

Engineer Battalion HHC Trainer Requirements

Position	SME	Rank/ Grade	MOS	# per Unit	# of Units per Site	Required # per Site
CO	*	CPT	EN	1	1	1
XO/1SG	*	E8	12Z	1	1	1
Support	*	E7	12B	1	1	1
Supply	*	E7	76Y/92Y	1	1	1
Total						4

Table C.40

Engineer Company Trainer Requirements

Position	SME	Rank/ Grade	MOS	# per Unit	# of Units per Site	Required # per Site
CO	*	CPT	EN	1	3	3
XO/1SG	*	CPT/E8	EN/12Z	1	3	3
Plt	*	E7	12B	3	3	9
Asst PLT	*	E7/E6	12B	3	3	9
Total						24

NOTE: Asst Platoon O/C-Ts observe and provide training feedback for squads and special equipment during lanes and battalion task force and brigade operations.

Table C.41

Engineer Battalion NCS Requirements

Position	SME	Rank/ Grade	MOS	# per Unit	# of Units per Site	Required # per Site
OIC	*	MAJ	EN	2	1	2
NCOIC	*	E8	12Z	2	1	2
Cmd., Ops, Intel	*	E7	12B	2	1	2
O/C-T	*	E7	12B	2	1	2
Driver		E4	12B	2	1	2
Total						10

Additional Gunnery and Lane Trainers, Managers, and Support Personnel

There is no specific allocation of personnel to support CEV gunnery. The events will be covered and managed by permanent trainers.

Table C.42

**Engineer Battalion Platoon Lanes Management and Support Requirements
(Days 27–43)**

Position	SME	Rank/ Grade	MOS	# per Lane	# of Lanes per Site	# per Site	# Required to Source
Lane Management PLT (Days 27–43)							
OIC (Note 1)	*	MAJ/CPT	EN	1	3	3	Covered by
NCOIC	*	E8/E7	12B	1	3	3	CO Lane Mgt.
Lane Support PLT (Days 27–43)							
Smoke Gen (2)	*	E5	54B	1	1	1	1
	*	E4	54B	1	1	1	1
Drivers/RTO		E5	12B	1	3	3	3
General		E4	12B	2	3	6	6
Lane and Support Total							11

NOTES: OIC and NCOIC operate NCS. Smoke generator only required on breach and remove obstacle lanes.

Table C.43

**Company Team Lanes Management and Support Requirements
(Days 42–50)**

Position	SME	Rank/ Grade	MOS	# per Lane	# of Lanes per Site	# per Site	# Required to Source
CO Management							
OIC	*	MAJ/CPT	EN	1	3	3	3
NCOIC	*	E7	12B	1	3	3	3
Net Control Sta.	*	E7/E6	12B	1	3	3	3
Total							9
CO Support							
Drivers/RTO		E5	12B	1	3	3	Covered by
General		E4		2	3	6	PLT Lane Supp
Company Team Lane Total							9
Engineer Total							72

AIR DEFENSE ARTILLERY (ADA) BATTERY TRAINING PERSONNEL

This outline of the ADA battery trainers is based on the current proposed organizational design of this battery. It has two BSFV platoons and two Stinger platoons and a headquarters. This organization has not been fully fielded, and modification of the trainer organization may be needed.

Permanent Training Personnel

These trainers align with the battery commander, first sergeant and executive officer, and BSFV and Stinger platoons. They provide training feedback and advice. They provide a permanent set of trainers for the battery and platoons. They also assist with control training and enforcement of rules of engagement. The Stinger platoon trainers conduct Stinger gunnery training and the BSFV platoon trainers support BFV gunnery training, when the BSFVs go through this training with the Mechanized Infantry battalion. During preliminary command and control training the platoon and company trainers support the TF and Brigade ADA trainers.

Table C.44

ADA Battery Trainer Requirements

Position	SME	Rank/ Grade	MOS	# per Unit	# of Units per Site	Required # per Site
Battery Trainer	*	CPT	ADA	1	1	1
XO/1SG	*	CPT/E8	ADA/14R or 14S	1	1	1
PLT/BSFV	*	E7	14R	2	1	2
PLT/STINGER	*	E7	14S	2	2	1
Total						6

FORWARD SUPPORT BATTALION (FSB) TRAINING TEAM

This organization was modeled on the NTC Operations Group's FSB Trainers or "Goldminers." It has been augmented to provide additional section and individual training capability. There are no lanes or gunnery augmentation personnel requirements. The personnel from this team would work with their counterparts from the other battalions of the brigade to conduct "CSS Lanes." It consists of the following groups of permanent training personnel.

- **FSB command and control trainers.** These trainers align with the battalion commander and the battalion's primary and key special staff sections. Their primary responsibility is to provide training feedback and advice to their counterparts, starting with individual and section-level training through completion of final preparations for deployment. They also assist with training control and enforcement of rules of engagement. These trainers remain with the battalion for the duration of training.

- **FSB HHC trainers.** These trainers align with the HHC commander, first sergeant, and executive officer. They provide training feedback and advice.

- **FSB companies.** These trainers align with the company commander, first sergeant and executive officer, and key sections and platoons. They provide training feedback and advice. They provide a permanent set of trainers for each company and platoon. They also assist with control training and enforcement of rules of engagement.

- **FSB net control section (NCS).** This section monitors the activities on the brigade and divisional nets, tracks ongoing operations for the trainers, and provides data to support AARs and other training requirements. All NCS operate on a two-shift basis.

Table C.45

FSB Command and Control Trainer Requirements

Position	S M E	Rank/ Grade	MOS	# per Unit	# of Units per Site	Required # per Site
Senior Tnr	*	LTC	OD/QM/TR	1	1	1
Sr. NCO	*	E9	91Z	1	1	1
S1	*	CPT	AG/O3A	1	1	1
SIDPERS	*	E7	75Z	1	1	1
S3	*	MAJ	QM/OD/TR	1	1	1
Ops NCO	*	E8/E7	63Z	1	1	1
S4	*	CPT	QM/OD/TR	1	1	1
Asst. S4	*	E8	76Z	1	1	1
NBC	*	E7	54B	1	1	1
Signal	*	CPT	SIG	1	1	1
BMO	*	CPT	OD	1	1	1
MMC / DMMC	*	WO/E8	915D/76Z	1	1	1
Total						12

NOTE: During C2 simulations, additional training support personnel are needed to operate the simulations. These would be provided by the installation or from RC personnel from the BCST of the USAR's Divisions Exercise, but are not included in our requirements.

Table C.46

FSB HHC Trainer Requirements

Position	S M E	Rank/ Grade	MOS	# per Unit	# of Units per Site	Required # per Site
HHC O/C-Ts						
CO	*	CPT	QM/OD/TR	1	1	1
XO/1SG	*	E8	63Z	1	1	1
Total						2

Table C.47

FSB Company Trainer Requirements

Position	S M E	Rank/ Grade	MOS	# per Unit	# of Units per Site	Required # per Site
A CO O/C-Ts						
CO	*	CPT	QM/TR/92B	1	1	1
XO/1SG	*	E8	88Z	1	1	1
Supply PLT	*	E8/E7	76Z	1	1	1
AMMO	*	E7	55B	1	1	1
POL	*	E7	77F	1	1	1
Total						5
B CO O/C-Ts						
CO	*	CPT	OD/91B	1	1	1
XO/1SG	*	E8	63Z	1	1	1
HQ PLT		WO/E7	915D/63H/912A	1	1	1
Shop	*	WO/E7	915D/63H	1	1	1
Tech Support	*	WO/E7	915D/63H/63E	2	1	2
Auto	*	WO/E7	915D/63H/63T	1	1	1
Total						7
C CO O/C-Ts						
CO	*	CPT	MSC	1	1	1
XO/1SG	*	E8	91B	1	1	1
HQ PLT	*	CPT/E7	91B/70B	1	1	1
Treat. PLT	*	E7	91B	2	1	2
Ambul. PLT	*	E7	91B	2	1	2
Total						7
Company Totals						19

Table C.48

FSB NCS Requirements

Position	SME	Rank/ Grade	MOS	# per Unit	# of Units per Site	Required # per Site
OIC	*	MAJ	QM/OD	2	1	2
NCOIC	*	E8	91Z/76Z	2	1	2
BN/BDE Cmd.	*	E7	11M/19K	2	1	2
BDE Admin/Log	*	E7	76Y/63H	2	1	2
Drivers		E4		2	1	2
Total						10
FSB Totals						43

CAVALRY TROOP TRAINING TEAM

The specific data we have been able to draw on for cavalry troop training requirements is limited. The requirements below have been derived from maneuver platoon and company requirements. There is no gunnery training requirement listed, as the M1s and M2 of the troop would conduct gunnery with the tank and mechanized infantry battalion. FIST and mortar trainers are shown in the FA battalion and maneuver battalion teams.

Table C.49

Cavalry Troop Trainer Requirements

Position	S M E	Rank/ Grade	MOS	# per Unit	# of Units per Site	Required # per Site
TRP	*	CPT	AR	1	1	1
XO/1SG	*	CPT/E8	AR/19Z	1	1	1
PLT	*	E7	19D (M3)	2	1	2
PLT	*	E7	19K (M1)	2	1	2
SEC/Dismount	*	E7/E6	19D (M3)	4	1	4
Total						10

Table C.50

Cavalry Troop Lanes Management and Support Requirements

Position	S M E	Rank/ Grade	MOS	# per Range	# of Ranges per Site	Required # per Site
PLT/TRP Lane Management (Days 13–66)						
OIC	*	MAJ/CPT	AR	1	1	1
NCOIC	*	E7/E6	19D	1	1	1
Net Control Sta	*	E7/E6	19D	2	1	2
Total						4
PLT/TRP Lane Support (Days 13–66)						
Fire Markers		E6/E5	19D	2	1	2
Drivers/RTO		E6/E5	19D	1	1	1
General		E4		2	1	2
Total						5
Lane Support and Management Total						9
Cavalry Troop Total						19

INDIVIDUAL WEAPONS AND DEMOLITION TRAINING

Table C.51

Individual Weapon and Demolition Trainer Requirements

Position	S M E	Rank/ Grade	MOS	# per Range	# of Ranges per Site	# per Site	# Required to Source
Range OICs/NCOICs and Trainers							
Small Arms—All (Days 17–53)							
M-16	*	E7/E6	11M/11B	2	1	2	2
M-9	*	E7/E6	11M/19D/12B	2	1	2	2
M-19	*	E7/E6	11M/11B	2	1	2	2
M249 / M-2	*	E7/E6	11M	2	1	2	2
AT-4	*	E7/E6	11M/11B	2	1	2	2
Total							10
Inf. & Engr. Squads (Days 17–47)							
Dragon	*	E7/E6	11M/11B	3	1	3	3
Hand grenade	*	E7/E6	11M/11B/11C	2	1	2	2
Demolition	*	E7/E6	12B	2	1	2	Covered by Permanent EN O/C-Ts
Total							5
Engr. Squads only (Days 21–25)							
Cratering charge	*	E7/E6	12B	2	1	2	Covered by Permanent EN O/C-Ts
Shape charge	*	E7/E6	12B	2	1	2	
Bangalore	*	E7/E6	12B	2	1	2	
Total							
Individual Weapons Total							15

NOTES: One unit provides all other scoring, support, and control. Two sufficient O/C-Ts from Engr. PLT lanes to support 12B requirements.

TRAINING PERSONNEL REQUIREMENTS AND SOURCES FOR THREE BRIGADE SITES

This appendix expands on the discussion in Chapter Three and details how the resources are provided to meet the requirements for three brigade-level training sites, each provided with the training personnel shown in Appendix C. It also details the specific number of personnel shortages by grade and by branch/MOS that could not be filled from available organizations.

While this appendix shows how the total training personnel requirement could be provided, it does not propose that the resources be actually allocated as described. We simply account here for the training personnel. For example, we will show all the personnel from the NTC Operations group as being at Fort Irwin, even though for actual implementation of three brigade sites it would probably be more beneficial to use the Operations Group to provide expertise to the other sites on a cadre basis. The exact design of the organization and providing it the personnel with the right qualifications will be crucial to the success of postmobilization training and will involve numerous considerations and tradeoffs, but such detail is beyond the scope of this study.

To calculate available personnel, we used authorized rank and branch/MOS strengths from the 1994 TDA of the NTC Operations Group, mounted warfare simulations trainers (MWST) at Fort Knox, the brigade command battle staff training (BCBST) team at Fort Leavenworth, readiness groups (RGs), and operational readiness evaluation (ORE) teams. We also used 1994 proposed TDAs for reserve training brigades (RTBs), resident training detachments (RTDs), and AC augmentation to the battle command staff training (BCST) brigades of Divisions Exercise.

Tables D.1, D.2, and D.3 detail the training personnel requirements and sources used to provide these requirements for each site. On the left side of each table is a summary of the requirements described in Appendix C. Across the top are the sources chosen to resource each site. The numbers in the table reflect the number used from each source to meet the requirement. The last column lists the needed training personnel that could not be provided because the required soldier of the required MOS or branch was not available from accessible sources.

Table D.1

Sources for Training Personnel for Site 1

Site 1	Total Req.	RTD	RTB AR	RTB IN	RTB EN	RTB FA	RTB HQ	BCST AC	ORE Teams	ARNG O/C-Ts	ARNG Tng Spt	Need
Support for mvr bns	135		19	8							108	0
AR bn	133	7	83	0				1	8		28	6
IN bns (2)	242	14		107	2				19	30	54	16
Engr bn	72	7			41				7		11	6
Mortar plts	15	0	4	4					1	1	5	0
Scout plts	14	0	1	11							2	0
CAV trp	19	2	9	4							4	0
FA bn + FISTs	99	7				58		1	11		13	9
Fwd spt bn	43	8							20		2	13
Bde HQs	43	3		1			3	21	2		4	9
Indiv/crew weapon	15	0		11	4						0	0
O/C-T HQ	137					1	37				99	0
	967	48	116	146	47	59	40	23	68	31	330	59

NOTES: The requirements for Lanes/Gunnery Support Team include the management and support personnel needed for gunnery ranges and company lanes.

Table D.2

Sources for Training Personnel for Site 2

Site 2	Total Req.	RTD	RTB AR	RTB IN	RTB EN	RTB FA	RTB HQ	BCST AC	ORE Teams	ARNG O/C-Ts	ARNG Tng Spt	Need
Support for mvr bns	135		19	8							108	0
AR bn	133	7	83	0				1	8		28	6
IN bns (2)	242	14		105	2			1	20	31	54	15
Engr bn	72	7			41				7		11	6
Mortar plts	15	0	4	4						1	5	1
Scout plts	14	0	1	11							2	0
CAV trp	19	2	9	4							4	0
FA bn + FISTs	99	7				57		2	8		13	12
FSB	43	8						2	16		2	15
Bde HQs	43	3		1			3	21	2		4	9
Indiv/crew weapon	15	0		11	4						0	0
O/C-T HQ	137					1	37				99	0
	967	48	116	144	47	58	40	27	61	32	330	64

Table D.3

Sources for Training Personnel for Site 3: Fort Irwin

Brigade Units	Total Req.	RTD	NTC Broncos	NTC Cobras	NTC Dragons	NTC Gold-miners	NTC Lizards & Outlaws	NTC Scorpions	NTC Sidewinders	NTC Tarantulas	NTC Were-wolves	NTC OPFOR	RTB IN	RTB AR	RTB FA	MWST	ORE Teams	ARNG Training Support	Need
All mnvr. bns—support	135	4		23	44													64	0
AR bn	133	6		11	7			27		1				27		21	4	28	1
IN bns (2)	242	10		21	31			20		15		70		14			7	48	6
Engr bn	62	5		5	9			4	25	4							2	7	1
Mortar plts	15			2	2			2		4									0
Scout plts	14				5			4		1				4				5	0
CAV trp	19	2												15				2	0
FA bn + FISTs	77	6	1	1	5			4		1	44				2	1	3	9	0
Forward support bn	33	7				15													9
Bde HQs	23	2	19														1		1
Indiv/crew weapon	15									12		2	1						0
O/C-T HQ	59						59												0
Tactical analysis facility	62																		
Total	889	42	20	63	103	15	59	61	25	38	44	72	1	60	2	22	19	163	18

NOTE: See Legend, next page.

Legend

NTC Operations Group
    Broncos: Brigade Staff
    Cobras: Armor Team
    Dragons: Live Fire Team
    Goldminers: Forward Support Battalion Team
    Lizards: Training Plans HQ Staff
    Outlaws: Training Operations HQ Staff
    Scorpions: Mechanized Infantry Team
    Sidewinders: Engineer Team
    Tarantulas: Light Infantry Team (some Heavy)
    Werewolves: Field Artillery Team
    OPFOR: Opposing Forces Teams
Regional Training Battalions (RTB)
    RTB AR: Tank battalion
    RTB IN: Infantry Battalion
    RTB EN: Engineer Battalion
    RTB FA: Field Artillery Battalion
    RTB HQ: Headquarters
BCST: Battalion Command Staff Training Brigade
RTD: Resident Training Detachment
MWST: Mounted Warfare SIM Trainers (Fort Knox)
ORE Teams: Operational Readiness Exercise Teams
ARNG O/C-Ts: National Guard

The comparison of requirements versus personnel in available organizations showed specific shortfalls. The shortages by grade and branch/MOS for a three-site requirement are shown on Table D.4. This list does not include shortages for the requirement to train the air defense artillery battery and the mechanized infantry company of the brigade at this time.

Table D.4

Shortages for a Three-Brigade Site

Rank	Branch/MOS	Number Short for Three Sites
CPT	Signal	3
CPT	ORD	2
CPT	MSC	6
CPT	MI	7
CPT	IN	4
CPT	CML	2
CPT	ADA	3
E7/8/9	91B	33
E8	88Z	2
E7	82C	2
E7/8	77F	2
E8	63Z	6
E7	54B	16
E7	31W	2
E7	19K	2
E7/6	13F	19
E7/WO	13R/131A	2
E8	12Z	4
E9/8	11Z	4
E8	11M	9
E7/6	11C	1
Total		141

NOTE: The total of 141 consists of 8 officers and 103 non-commissioned officers. This does not include the requirement for ADA battery trainers, which we estimate to be one officer and five NCOs.

GARRISON SUPPORT REQUIREMENT

To determine what level of support might be required to sustain two brigades (one OPFOR and one enhanced) at an installation after mobilization, we reviewed the installation support at two-brigade active component installations. We chose Forts Carson, Riley, Stewart, and Irwin. We then determined what TDA and nondivisional TOE units were assigned there, and surveyed installation personnel to determine what elements they felt would be needed to support two brigades during an intensive training period. We told them to assume that their mobilization TDAs were present. Although this methodology has a strong subjective component, it does provide a way to approximate the garrison support element.

We assumed that the TDA units would not deploy but that TOE units would. TDA units included the garrison headquarters and the medical activities. TOE units included adjutant general, medical, military policy, engineer, quartermaster, transportation, aviation, ordnance, and signal.

The results of the analysis appear in Table E.1. The first three columns display the units and authorized strength of the three installations, and the fourth column shows the support the three installations considered necessary to support the enhanced brigades.

Table E.1
Garrison Support at Two-Brigade AC Installations

	Fort Carson		Fort Riley		Fort Stewart		Typical 2-Brigade Site	
	Unit Type	Auth	Unit Type	Auth	Unit Type	Auth	Unit Type	Auth
Permanent TDA								
	Garrison HQ	362	Garrison HQ	240	Garrison HQ	383		
	Medical Activity	438	Medical Activity	399	Medical Activity	474		
TOE Units								
AG								
	AG CO	129	AG CO	129	AG CO	214	AG CO	129
	Finance Bn	84	Finance Bn	65	Finance Bn	103	Finance Bn	84
	PA Team	5			PA Team	5	PA Team	5
					AG Replace. Det.	30		
Medical								
	Cmt. Hospital	465	MASH Hsp.	151			MASH Hsp.	151
	Prev. Det. (Sani)	10	Prev. Det. (Sani)	10	Prev. Det. (Sani)	10	Prev. Det. (Sani)	10
	Dental Det.	23			Dental Det.	23	Dental Det.	23
	Air Ambulance	130	Air Ambulance	130			Air Ambulance	130
	Stress Det.	21			Vet. Det (small)	6	*or Med. Helicopter CO	6
Military Police								
	Hvy. CID Det.	11	Hvy. CID Det.	11	Hvy. CID Det.	11	Hvy. CID Det.	11
	Cbt. Support CO	177	Cbt. Support CO	177	Cbt. Support CO	177	Cbt. Support CO	177
Engineers								
	Fire Truck Det.	6						
Supply								
	CS CO	134	CS CO	134	CS CO	177	CS CO	134

Table E.1—continued

Category	Fort Carson Unit Type	Auth	Fort Riley Unit Type	Auth	Fort Stewart Unit Type	Auth	Typical 2-Brigade Site Unit Type	Auth
Quartermaster					POL Supply BN	56		
					Petro. Sup. CO	199		
					Water Sup. CO	148		
					Water Pur. Det.	52		
Transportation	Med. Truck CO	140	Med. Truck CO	140	Lt./Med. Truck CO	117	Lt./Med. Truck CO	117
	Med. Truck 7.5 G	164			Lt./Med. Truck CO	117		
	TC Mvt. Con. Det.	7	TC Mvt. Con. Det.	5	TC Mvt. Con. Det.	7	TC Mvt. Con. Det.	7
	TC Mvt. Con. Det.	4	TC Mvt. Con. Det.	3				
Aviation	Med. Helo. CO	176			Avn. Reg. CO	248		
Ordinance	EOD Team	17	EOD Team	17	EOD Team	23	EOD Team	17
			Ammo. CO	175	Ammo. CO	175		
	Non-Div Maint. CO	235	CS Maint. CO	256	Non-Div Maint. CO	242	Non-Div Maint. CO	235
	GS Maint. CO	218	CS Maint. CO	211	CS Maint. CO	296	GS Maint. CO	218
	ATE Repair Det.	7	ATE Repair Det.	7	ATE Repair Det.	13	ATE Repair Det.	7
DIV Signal	MSE BN (4 nodes)	533	MSE BN (4 nodes)	517	MSE BN (6 nodes)	677	MSE BN (2 nodes)	280
TOTAL TOE		2,696		2,004		3,126		1,735

OPFOR REQUIREMENTS

This appendix expands on Chapter Five and describes in greater detail the OPFOR requirements to support execution of the training model.

OPFOR ORGANIZATION

The OPFOR organization we used was modeled on the NTC's Krasnovian threat, basically a modified Soviet-styled motorized rifle regiment (MRR).[1] The MRR has three motorized rifle battalions (MRB) and one tank battalion. At the NTC, the tank battalion's tanks are assigned to the MRR to form reinforced rifle elements. Table F.1 shows the composition of tanks, IFVs, other fighting vehicles, and dismounted infantry in these reinforced motorized rifle elements.

Table F.1

Reinforced Motorized Rifle Units

Unit	Elements	Tanks	IFVs	Other Combat Vehicles	Dismounted Infantry
Reinforced motorized rifle platoon (MRP)		1	3		27
Reinforced motorized rifle company (MRC+)	3 MRP and HQ Tank and IFV	4	10		63
Reinforced motorized rifle battalion (MRB+)	3 MRC and HQ Tank	13	120	5[a]	210[b]
Total		39	120	5	

[a]Other combat vehicles are 5 command control vehicles for the anti-tank platoon and battalion commander.

[b]A total of 27 7-man rifle squads and a 21-man grenade launcher platoon.

---

[1]See Army Field Manual FM 100-2-3, *The Soviet Army*, July 1984, for greater detail on the exact organization of a MRR.

Below we list other MRR elements portrayed during the training events in our training model, along with their number of combat vehicles.

Unit	Combat Vehicles
1 howitzer battalion	18 122mm SP
1 air defense artillery battery	4 SA 9, 4 ZSU-23-4, 3 BTR 60
1 anti-tank battery	9 AT 5, 4 BDRM
1 engineer company	6 ACE, 6 CEV, 9 AVLB, 8 M113[a]
1 reg reconnaissance company	4 BMP, 4 BDRM

[a]These are U.S. vehicles. They are the closest counterparts to equivalent Soviet vehicles.

OPFOR SUPPORT FOR COMPANY TEAM LANES

There are nine company team lanes in the training model. Table F.2 shows the breakdown of these lanes and the OPFOR requirement.

There is also a requirement to support self-defense training of other units that occurs concurrently with company lanes, including *spetnatz*-type activities. OPFOR support for these events is approximately 70 dismounted infantry. We felt that this additional requirement could be met by personnel performing primary OPFOR duties in the company lanes.

Table F.2

OPFOR Requirements for Company Lanes

Lane	OPFOR Element	Tanks	BMP	Other Combat Vehicles	Dismounted Infantry
Hasty defense	2 MRC(+)	8	20	0	126
Reposition	1 MRC(+/–)	4	10	2	0
Deliberate defense	2 MRC(+)	8	20	3	126
Counterreconnaissance	Rgt recon		4	4	9
Movement to contact	MRP(+)	4	10	0	63
Breach	MRP(+)	2	3	0	21
Assault	MRP(+)	1	3	0	21
Advance guard/support by fire	MRC(+)	4	10	3	21
Trench	MRP(+)	1	3	1	21
Cavalry troop[a]	MRC(+)	4	14	7	21
Total		36	97	20	429[b]

NOTE: MRC = motorized rifle company; MRP = motorized rifle platoon.

[a]The cavalry troop lanes are run on five special lanes and two lanes that maneuver company teams execute. The numbers shown are the high number required for the special lanes.

[b]Mainly reconnaissance company and engineer company requirements.

BATTALION AND BRIGADE OPERATIONS

A BLUEFOR battalion task force defense in sector requires a reinforced OPFOR motorized rifle regiment, or MRR(+). BLUEFOR battalion task force deliberate attack and movement to contact require a reinforced motorized rifle battalion, or MRB(+). The BLUEFOR brigade deliberate attack requires much of an MRR(+). Thus the schedule in our training model requires one MRR(+) and one MRB(+) for each training site. The OPFOR requirement for one site is shown in Table F.3.

Table F.3

OPFOR Requirement for Battalion and Brigade Operations

	Tanks	BMPs	Other Combat Vehicles	Dismounted Infantry
MRB(+)	13	40	26	210[a]
MRR(+)	40	132	63	220[a]
Total	53·	172	89[b]	430

[a]These are actually considerably below the actual dismounted strength for these units. An MRB has 189 dismounted infantry in the assigned rifle squads, and an MRR has 567. To portray an MRR and MRB, the theoretical requirements would be over 800 if all rifle squads, dismounted scouts, and weapon systems were included. However, this would require almost two full mechanized infantry brigades. For purposes of achieving the desired training objectives, we have developed a restrained requirement—one that can be supported by an ARNG mechanized brigade. For offensive operations, we have allocated far less than the actual requirement. The numbers above provide enough to portray reasonably full OPFOR defensive arrays and provide enough OPFOR dismounted strength for OPFOR offensive operations to train BLUEFOR defensive tasks. Achieving sufficient dismounted strength levels could be difficult, given possible low strength levels in many lower-priority RC units. Preparing for many likely future threats (e.g., Korea) would require scenarios and OPFOR with more emphasis on dismounted operations than currently occurs at the NTC.

[b]This again is somewhat less than the actual number but sufficient to meet training objectives.

General

Aspin, Les, *Report on the Bottom-Up Review*, Washington, D.C.: Department of Defense, October 1993.

BCBST, briefing, Combined Arms Command, Fort Leavenworth, Kansas.

Department of the Army Inspector General (DAIG), *Special Assessment: National Guard Brigades' Mobilization*, Washington, D.C., June 1991.

General Accounting Office, *National Guard: Peacetime Training Did Not Prepare Combat Brigades for the Gulf War*, General Accounting Office, Report GAO/NSIAD-91-263, Washington, D.C., 1991.

General Accounting Office, *National Guard: Peacetime Training Did Not Prepare Combat Brigades for the Gulf War*, General Accounting Office, Report GAO/NSIAD-92-67, Washington, D.C., 1992a.

General Accounting Office, *Operation Desert Storm: Army Had Difficulty Providing Adequate Active and Reserve Support Forces.* General Accounting Office, Report GAO/NSIAD-93-4, Washington, D.C., 1992b.

Grossman, Jon, *Battalion Level Command and Control at the National Training Center*, Santa Monica, Calif.: RAND, MR-496-A, 1994.

Headquarters Department of the Army, *Total Army Basing Study*, Washington, D.C., January 1993.

———, DAMO-TRR, *Report of the Enhanced Brigade Task Force*, April 1994.

Lee, Deborah R., *Prepared Statement of the Honorable Deborah R. Lee, Assistant Secretary of Defense for Reserve Affairs, Before the National Security Subcommittee, House Appropriations Committee*, March 14, 1995.

Lippiatt, Thomas, F., J. Michael Polich, and Ronald E. Sortor, *Post-Mobilization Training of Army Reserve Component Units*, Santa Monica, CA: RAND, MR-124-A, 1992.

Sortor, Ronald E., Thomas F. Lippiatt, J. Michael Polich, and James C. Crowley, *Training Readiness in the Army Reserve Components*, RAND, Santa Monica, California, MR-474-A, 1994.

Army Training Documents

1st Cavalry Division, *155th Brigade Annual Training Plan*, briefing, Fort Hood, Texas, June 1993.

———, *155th Brigade Post-Mobilization Training Plans*, Fort Hood, Texas, unpublished, 1994.

1st Infantry Division (Mech), *218th Brigade Post-Mobilization Training Plans*, Fort Riley, Kansas, unpublished, 1994.

2nd Armor Division, *256th Brigade Annual Training Plan*, briefing, Fort Hood, Texas, June 1993.

———, *256th Brigade Post-Mobilization Training Plans*, Fort Hood, Texas, unpublished, 1994.

4th Infantry Division (Mech), *After Action Report on 155th Armor Brigade Mobilization and Training*, Fort Carson, Colorado, unpublished, April 1991.

———, *Iron Point Lane Training Program*, Fort Carson, Colorado, July 1992.

———, *116th Brigade Annual Training Plan*, briefing, Fort Carson, Colorado, June 1993.

———, *116th Brigade Post-Mobilization Training Plans*, Fort Carson, Colorado, unpublished, 1994.

5th Infantry Division (Mech), *After Action Report on Training of 256th Infantry Brigade During ODS*, Fort Polk, Louisiana, unpublished, March 1991a.

———, *256th Brigade Training in ODS*, briefing, Fort Polk, Louisiana, March 1991b.

24th Infantry Division, *48th Brigade Annual Training Plan*, briefing, Fort Stewart, Georgia, June 1992.

Department of the Army, Training Circular 25-1, *Training Land*, Washington, D.C., September 1991.

———, ARTEP 5-145-DRILL, *Engineer Drills*, Washington, D.C., October 1990.

———, ARTEP 7-7J-DRILL, *Battle Drills for the Bradley Infantry Fighting Vehicle Platoon, Section and Squad*, Washington, D.C., December 1992.

———, Mission Training Plan ARTEP 44-177-30-MTP, *Mission Training Plan for an ADA Battery Bradley Stinger Fighting Vehicle*, Washington, D.C., September 1994.

———, Mission Training Plan ARTEP 44-117-21-MTP, *Mission Training Plan for Avenger Platoon*, Washington, D.C., June 1992.

———, Mission Training Plan ARTEP 17-57-10-MTP, *Mission Training Plan for Scout Platoon*, Washington, D.C., December 1988.

———, Mission Training Plan ARTEP 42-077-30-MTP, *Mission Training Plan for Supply and Transport Company, Support Battalion, Heavy Separate Brigade or Infantry Separate and Supply and Transport Troop, Support Squadron, Armored Cavalry Squadron*, Washington, D.C., May 1991.

———, Mission Training Plan ARTEP 5-145-31-MTP, *Mission Training Plan for the Combat Engineer Company, Heavy Division/Corps/Armored Cavalry Regiment*, Washington, D.C., September 1990.

———, Mission Training Plan ARTEP 5-145-11-MTP, *Mission Training Plan for the Combat Engineer Platoon, Heavy Division/Corps/Armored Cavalry Regiment*, Washington, D.C., February 1989.

———, Mission Training Plan ARTEP 6-115-30-MTP, *Mission Training Plan for the Field Artillery Cannon Firing Battery*, 155mm Self Propelled, Washington, D.C., January 1990.

———, Mission Training Plan ARTEP 6-115-20-MTP, *Mission Training Plan for the Field Artillery Fire Support Battalion* Washington, D.C., January 1990.

———, Mission Training Plan ARTEP 6-115-31-MTP, *Mission Training Plan for the Field Artillery—Cannon, Battalion Headquarters and Headquarters Battery; Headquarters and Headquarters Battery and Service Battery*, Washington, D.C., November 1990.

———, Mission Training Plan ARTEP 5-145-MTP, *Mission Training Plan for the Headquarters and Headquarters Company, Engineer Battalion, Heavy Division/Corps*, Washington, D.C., February 1989.

———, Mission Training Plan ARTEP 63-085-MTP, *Mission Training Plan for the Headquarters, Support Battalion, Separate Infantry Brigade and Heavy Separate Brigade*, Washington, D.C., December 1990.

———, Mission Training Plan ARTEP 71-3-MTP, *Mission Training Plan for the Heavy Brigade Command Group and Staff*, Washington, D.C., October 1988.

————, Mission Training Plan ARTEP 7-90-MTP, *Mission Training Plan for the Infantry Mortar Platoon, Section and Squad*, Washington, D.C., August 1989.

————, Mission Training Plan ARTEP 7-8-MTP, *Mission Training Plan for the Infantry Rifle Platoon and Squad*, Washington, D.C.

————, Mission Training Plan ARTEP 8-437-30-MTP, *Mission Training Plan for the Medical Company. Support Battalion. Heavy Separate Brigade/Separate Infantry Brigade and Medical Troop, Support, Squadron, Armored Cavalry Regiment*, Washington, D.C., September 1993.

————, Mission Training Plan ARTEP 43-079-30-MTP, *Mission Training Plan for the Ordinance (Maintenance) Company, Support Battalion, Separate Infantry Brigade or Heavy Separate Brigade*, Washington, D.C., June 1991.

————, Mission Training Plan ARTEP-MTP 17-487-30-MTP, *Mission Training Plan for the Regimental Armored Cavalry Troop*, Washington, D.C., September 1991.

————, Mission Training Plan ARTEP 71-2-MTP, *Mission Training Plan for the Tank and Mechanized Infantry Battalion Task Force*, Washington, D.C., October 1988.

————, Mission Training Plan ARTEP 71-1-MTP, *Mission Training Plan for the Tank and Mechanized Infantry Company Team*, Washington, D.C., October 1988.

————, Mission Training Plan ARTEP-MTP 17-237-10-MTP, *Mission Training Plan for the Tank Platoon*, Washington, D.C., October 1988.

————, Mission Training Plan ARTEP-MTP 17-236-10-MTP, *Mission Training Plan for the Task Force Maintenance Platoon*, Washington, D.C., December 1987.

————, Mission Training Plan ARTEP-MTP 17-236-12-MTP, *Mission Training Plan for the Task Force Medical Platoon*, Washington, D.C., December 1987.

————, Mission Training Plan ARTEP-MTP 17-236-11-MTP, *Mission Training Plan for the Task Force Support Platoon*, Washington, D.C., November 1987.

————, Pamphlet 350-38, *Standards in Weapons Training*, Washington, D.C., February 1993.

NTC Operations Group, *48th Infantry Brigade Lanes Training Packages*, National Training Center, Fort Irwin, California, unpublished, 1990a.

————, *48th Infantry Brigade ODS Training Plans*, National Training Center, Fort Irwin, California, unpublished, 1990b.

————, *NTC Rotation 94-08 Schedule*, National Training Center, Fort Irwin, California, 1994a.

———, *Personnel Allowance*, National Training Center, Fort Irwin, California, January 1994b.

FORSCOM Documents

U.S. Army Forces Command, *Active Component Staffing for Battle Command Staff Training (BCST) Brigades*, Fort McPherson, Georgia, September 1994.

———, *Active Component to Reserve Component Support*, briefing, Fort McPherson, Georgia, March 1995.

———, *Armor School TDA Personnel Allowance for Mounted Warfare SIM Trainers*, Fort Knox, Kentucky, June 1994.

———, *Fifth U.S. Army Active Component Field Exercise Brigade*, Fort McPherson, Georgia, September 1994.

———, *Fifth U.S. Army Regional Training Brigade Personnel*, Fort McPherson, Georgia, April 1995.

———, *Fifth U.S. Army Reserve Component Support Team TDA Personnel Allowances*, Fort McPherson, Georgia, March 1994.

———, *First U.S. Army Active Component Field Exercise Brigade*, Fort McPherson, Georgia, September 1994.

———, *First U.S. Army Regional Training Brigade Personnel*, Fort McPherson, Georgia, April 1995.

———, *First U.S. Army Reserve Component Support Team TDA Personnel Allowances*, Fort McPherson, Georgia, March 1994.

———, *Ground Force Readiness Enhancement OPPLAN, Draft*, Fort McPherson, Georgia, December 1994.

———, *Mobilization Station Study*, Fort McPherson, Georgia, February 1994.

———, *Regulation 350-2*, Fort McPherson, Georgia, June 1995.

———, *Resident Training Detachment Personnel Allowance*, Fort McPherson, Georgia, July 1994.

———, *Second U.S. Army Active Component Field Exercise Brigade*, Fort McPherson, Georgia, September 1994.

———, *Second U.S. Army Regional Training Brigade Personnel*, Fort McPherson, Georgia, April 1995.

———, *Second U.S. Army Reserve Component Support Team TAA*, March 1994.

————, *Sixth U.S. Army Active Component Field Exercise Brigade*, Fort McPherson, Georgia, September 1994.

————, *Sixth U.S. Army Regional Training Brigade Personnel*, Fort McPherson, Georgia, April 1995.

————, *Sixth U.S. Army Reserve Component Support Team TDA Personnel Allowances*, Fort McPherson, Georgia, March 1994.

————, *Training Assessment Model (TAM)*, U.S. Army Forces Command, Fort McPherson, Georgia, May 1993.

Army Field Manuals

Department of the Army, Field Manual FM 100-2-3, *The Soviet Army*, Washington, D.C., July 1984.

————, Field Manual FM 25-100, *Training the Force*, Washington, D.C., September 1988.

————, Field Manual FM 25-101, *Battle Focused Training*, Washington, D.C., September 1990.

————, Field Manual FM 100-5, *Battle Operations*, Washington, D.C., September 1990.

————, Field Manual FM 23-1, *Bradley Fighting Vehicle Gunnery*, Washington, D.C., September 1990.

————, Field Manual FM 17-12-1, *Tank Combat Tables, M1*, Washington, D.C., September 1990.

Maps

555th Engineer Company, *Fort Hood Military Installation Map*, Fort Hood, Texas, 1976.

Defense Mapping Agency Hydrographic/Topographic Center, *Fort Irwin Military Installation Map*, Bethesda, Maryland, 1992.

————, *Fort Riley Military Installation Map*, Washington, D.C., 1981.

————, *Fort Stewart Military Installation Map*, Bethesda, Maryland, 1992.

————, *Piñon Canyon Maneuver Site*, Washington, D.C., 1984.